IN REFEREES WE TRUST?

The MIT Press
Massachusetts Institute of Technology
77 Massachusetts Avenue
Cambridge, MA 02139
mitpress.mit.edu

The Open Access edition of this book was made possible by generous funding and support from MIT Press Direct to Open and a grant from the Alfred P. Sloan Foundation.

The MIT Press would like to thank the anonymous peer reviewers who provided comments on drafts of this book. The generous work of academic experts is essential for establishing the authority and quality of our publications. We acknowledge with gratitude the contributions of these otherwise uncredited readers.

This book was set in Stone Serif and Stone Sans by Westchester Publishing Services. Printed and bound in the United States of America.

Library of Congress Cataloging-in-Publication Data is available.

ISBN: 978-0-262-05474-4

EU Authorised Representative: Easy Access System Europe, Mustamäe tee 50, 10621 Tallinn, Estonia | Email: gpsr.requests@easproject.com

To all the referees who have been Reviewer 1.

You may not be the subject of memes, but your positive feedback is appreciated.

CONTENTS

ACKNOWLEDGMENTS

Given the subject of this book, it feels appropriate to begin by thanking the anonymous referees who read my proposal and manuscript for the MIT Press. My referee reports offered both encouragement and constructive criticism, and the referees' careful, rigorous engagement with my manuscript made this a better book. All of you made this process a model of what the peer review system can be when thoughtful scholars take their referee work seriously.

This book has been more than a decade in the making. To be clear, that has not been a decade of steady progress. I began working on this project while bouncing between short-term fellowships and adjunct positions, put it on the back burner when I left academia, and then had to keep it on the back burner for a while longer because my return to academia coincided with the first years of the COVID-19 pandemic. The following institutions sponsored my research leaves and enabled me to turn years of reading, archival visits, and thinking into actual written work: the American Council of Learned Societies, the University of Maryland's ISRCA program, the Alfred P. Sloan Foundation, and the Wilson Center in Washington, DC. The Wilson Center, a nonpartisan government think tank for policy scholarship, was defunded by the Trump administration during a purge of programs that supported academic research, and I am deeply saddened to know that future scholars may not have the same opportunity I did to work there.

Thank you to the fellow scholars who read parts of this book and offered their feedback on it, and who offered their mentorship and support as I worked to complete the manuscript: Michael Barany, Sarah Cameron, Alex Csiszar, Ryan Dahn, Michael Gordin, Evan Hepler-Smith, David Kaiser, Ahmet Karamustafa, Carole Lee, Bernard Lightman, Patrick McCray, Erika Milam, Tom Zeller, and all of the members of the Wilson Center 2024–2025 class of fellows. Special thanks to Aileen Fyfe, who was incredibly generous with her time and advice on this book. My Maryland colleague Antoine Borrut reached out to me as a new faculty member and gave me sound advice on applying for fellowships to allow me to devote time to this project. Lisa Klein patiently and expertly helped me navigate the financial side of applying for grant money; Meghann Babo read and gave me incredibly useful feedback on my funding proposals.

I also want to thank my colleagues at *Physics Today*, the magazine where I worked from 2016 to 2020. That job transformed my writing and gave me invaluable new perspectives on both modern science and the publishing world. To Steve Blau, Cindy Cummings, Charles Day, Rich Fitzgerald, Tonya Gary, Andrew Grant, Paul Guinnessy, Marty Hanna, Heather Hill, Alex Lopatka, Christine Middleton, Johanna Miller, Donna Padian, Freddie Pagani, Gayle Parraway, Ashley Smart, Rita Wehrenberg, and Mark Wilson: Thank you for your support, your kindness, your passion for science, and for everything you taught me about clear and concise writing. I suspect that, despite my best efforts, there are excess commas and dangling modifiers in this book. I take full responsibility for them.

This book has benefited from the work of two highly capable research assistants: Nora Zhang, who worked with me as a Wilson Center intern, and Ellen Adams, a UMD graduate student who worked with me as an RA for much of 2024 and 2025. You will see Nora's work reflected in the material drawn from the Columbia University Archives and the David Lazarus Papers, which she aided me in obtaining and processing. She also helped me locate and catalog media coverage of peer review in the 1970s and 1980s. Ellen's work was crucial to my second chapter; she helped me comb through and take notes on the scans I took at the National Academies of Science and the Rockefeller Foundation Archives. She also edited all of my footnotes to ensure they were complete. I would still be trying to write a first draft of chapter 2 if not for her efforts, to say nothing of the state of my citations.

Many thanks to Gita Manaktala, my editor at the wonderful MIT Press, whose enthusiasm and support have been key to getting this book over the finish line. Audra Wolfe of The Outside Reader worked with me at several stages of this project, from the proposal to the final draft; her keen editorial eye and historical expertise have been invaluable.

Finally, thank you to my family. My parents Linda and Bob and my in-laws Katherine and Mark supported me immeasurably as I worked to complete my revisions during the summer of 2025. My husband, Scott, has listened to me talk through writing snarls over dinner on more nights than I can count, and my sons, Hunter and Evan, are always ready to cheer me up with a silly dance (Evan) or a board game (Hunter; he usually wins). I can't wait to celebrate publication with you guys.

INTRODUCTION

Almost every academic can tell you the story of their first experience with peer review. Nearly twenty years later I still vividly remember mine. I was a third-year graduate student visiting Reading, England, to conduct archival research for my dissertation. It was 2008 and Wi-Fi was not yet standard at the kind of hotel that fit into a grad student budget, and so every day after my archives closed, I would walk to an internet café to check my email. One day I opened my inbox to find a message from the journal where I'd recently submitted my first scholarly paper. The editorial staff at *Notes and Records of the Royal Society of London* informed me that two anonymous historians of science had read my paper and written reports about it.[1] To my delight, the experts had deemed my work publishable provided I made some changes, ranging from making a more compelling argument about my paper's central figure to clarifying several points of potential confusion. One of the two referees noted that some of my phrases were "inappropriate" but remarked that this was "not unusual for an American author."

Although I was a little startled by the idea that I'd written something "inappropriate" (and I never did figure out which phrases the referee meant), nothing else about the process struck me as odd or surprising. As the daughter of a scientist and a graduate student in a research department, I fully expected that I would have to respond to feedback from anonymous experts before I could publish my work, or at least before I could publish it anywhere that counted for my career. I considered peer review an intrinsic part of scholarship, and it was obvious to me that peer-reviewed papers and journals should be considered, as one library guide puts it, "the most reliable information sources."[2] Even as I began work on a dissertation about the history of the scientific journal *Nature*, I didn't think to wonder about where the peer review process had come from or how it had become an expected part of scholarly publishing.

That began to change after I went to London to interview David Davies, the Welsh geophysicist who edited *Nature* from 1973 to 1980. *Nature* had, unfortunately, not preserved its editorial archives, and so talking to Davies was going to be my first and clearest look at how the editorial office was run in the late twentieth century. During our interview, Davies casually tossed out the information that when he arrived, one of his first jobs was "getting the refereeing system beyond reproach" and stopping the practice of publishing papers without referee reports.[3]

I was stunned. *Nature*, the journal known for publishing the theory of continental drift and the structure of DNA, did not require external peer review for all papers it published until *1973*? I'd long since learned that nineteenth-century *Nature* didn't have any system resembling modern peer review, but the idea that something could have been published without peer review in the 1970s felt unfathomable to me. I began to wonder if I'd just uncovered a scandal. But Davies didn't seem to think there was anything remarkable about what he'd said, and so when I got home, I began digging into information about refereeing and peer review in the late twentieth century. It turned out that I had not uncovered a historical bombshell at all. It was, in fact, extremely common for scholarly journals to referee only some papers, or for them to rely entirely on editorial judgments, well into the late twentieth century.

By the time I entered graduate school in the first decade of the twenty-first century, however, peer review was part of publishing in every scholarly field. Today, most of the search engines I use tell me whether a journal I am considering reading is peer reviewed (figure 0.1).[4] Grants all over the world assure applicants that their proposals will be peer reviewed by fellow experts. And passing the test of peer review seems especially important in the sciences, where non-peer-reviewed findings are often regarded

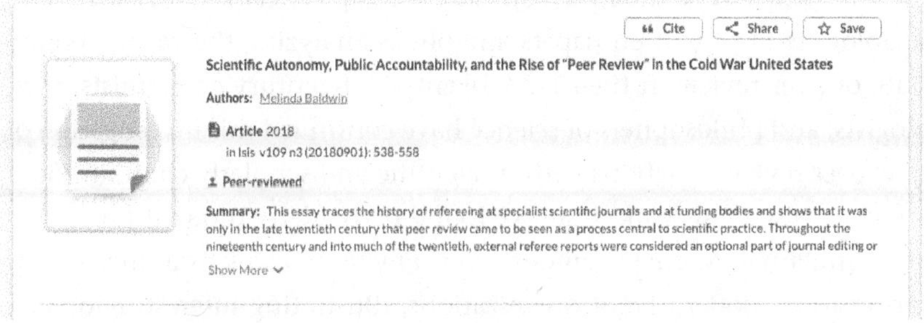

0.1 A January 2024 screenshot from the WorldCat library search engine. The search engine reassures potential readers that my 2018 article about peer review was, in fact, published in a peer-reviewed journal. *Source:* https://umaryland.on.worldcat.org/oclc/10115329566.

as unreliable or preliminary. In September 2012, when *Physics Letters B* accepted a paper about the Higgs boson, one outlet reported the event under the headline "CERN's Higgs Boson Discovery Passes Peer Review, Becomes Actual Science."[5] So how did we arrive at this point? When and why did "peer review" become synonymous with "actual science"?

In Referees We Trust? is an account of the rise of peer review in the sciences that seeks to understand how peer review came to be linked with scientific credibility. My central claim is that the idea that "peer review" is a prerequisite for scientific legitimacy solidified during conflicts about expert advice at federal funding bodies in the Cold War United States. We know that this idea emerged specifically in the US because, as we shall see, commentary from international scientists during the 1960s and 1970s identified "peer review" as a uniquely American obsession, one that many scientists outside the US were skeptical about. The importance of the US context inspired the title *In Referees We Trust?*, a nod to the phrase "In God We Trust" printed on US currency. Refereeing was not an American invention, as we will see in chapter 1, but it was the US scientific community that first turned peer review into a "cornerstone of science."[6]

BACKGROUND AND EXISTING SCHOLARSHIP

I am far from the first scholar to examine the role of peer review in the sciences. Because refereeing is so important to scientific and scholarly work, over the past several decades many scientists, physicians, journalists, and sociologists have written papers and books analyzing the contemporary state of peer review in their era.[7] Twenty-first-century economists, sociologists, and philosophers of science have continued this investigation of how peer review functions within scientific and scholarly communities.[8] To name just a few works that I've found particularly thought-provoking and influential, Michèle Lamont's *How Professors Think* examines review processes at modern grant organizations, illustrating interesting divides in what different fields consider "excellent" work and illuminating the social functions peer review serves for members of the scholarly community.[9] The philosopher David Shatz's 2004 book *Peer Review: A Critical Inquiry* examines the epistemology of refereeing, analyzing some common assumptions about peer review's function in academic journal publishing.[10] The philosopher Carole Lee and her collaborators have studied bias in peer review, particularly in the context of grants.[11] Some of these works offer a brief sketch of the history of refereeing, but a full account has yet to be written. I seek to extend the existing scholarship by offering a historical analysis of where systems of peer review came from, how they developed, and how ideas about what "peer review" is for have changed over time.

When I began my research, I thought of this project as a history of journal refereeing. Peer review at grants seemed to me to be an entirely different phenomenon—as John Burnham, one of the few historians who had written about peer review in the sciences at the time, put it, "Grant peer reviewing and editorial peer reviewing developed independently and, as far as I can determine at this point, had no influence on each other."[12] The two types of review make markedly different interventions in the scientific process and, to this day, take on noticeably different forms. Many funding bodies still gather panels of experts to review proposals in person; journals, in contrast, send papers out for reports from a small number of referees. While journal referees are, in theory, only judging the paper at hand, grant applicants have to submit not just a proposal explaining what they want to do with the money and why, but CVs and evidence of previous success. Grant referees explicitly consider the applicant's background as a factor in their decision, making the referee report a judgment about both this particular proposal and the scientist's body of work as a whole. There is also an epistemological difference. Journal referees are evaluating science that has already been performed and judging whether the paper's conclusions ought to be given the stamp of authenticity that publication in a specialist journal bestows. Grant referees are, in theory, judging science that has yet to be done, and there is often an implicit sense in discussions about grant refereeing that even the best proposals pose something of a risk.

And so initially I thought I would be on solid ground if I wrote a history of refereeing that focused only on journals. However, I quickly realized that if I wanted to explain how peer review came to be linked to scientific credibility, I would have to expand my scope. Burnham was right that the two systems had different origins and developed along separate tracks for much of their histories, but controversies about grant review were central to spreading the belief that "peer review" was a critical part of properly functioning science. I could not fully understand our modern ideas about the role and function of refereeing while focusing solely on journals. To understand where we got the idea that science has to be peer reviewed to be considered legitimate, my book would have to include grant review—and would have to pay particular attention to both the practice of, and the discourse surrounding, grant review at US federal funding bodies.

This book has benefited enormously from recent scholarly interest in the history of scientific periodicals. Books and articles on scientific publishing have illuminated the importance of editors, the ways readers interacted with specialist scientific texts, and the complex relationship between editorial policies and periodical credibility.[13] I recommend two books in particular to anyone interested in the history of journals: Alex Csiszar's *The Scientific Journal: Authorship and the Politics of Knowledge* and Aileen Fyfe, Noah Moxham, Julie McDougall-Waters, and Camilla Mørk Røstvik's *A History of Scientific Journals: Publishing at the Royal Society, 1665–2015.*[14] I have also been able to draw on excellent historical work about grants and funding bodies by scholars including Natalie Aviles, Michael Barany, Peter Huistra, Noortje Jacobs, Robert Kohler, Roy Macleod, Richard Mandel, George Mazuzan, Robin Wolfe Scheffler, and Laura Stark.[15]

The contribution of *In Referees We Trust?* is twofold. First, it is the first work of historical scholarship to trace the development of refereeing in the sciences over a long-term history, including both journals and funding bodies. Second, and more significantly, it explains the genesis and rise of the idea that peer review is a process that, as some recent papers have put it, "ensures scientific quality"[16] and is "crucial to science."[17] Neither of those claims would have been made by the earliest adopters of refereeing, but they are now commonplace, echoed everywhere from college science classrooms to government hearings to the scientific papers I quote throughout this introduction. That idea first began to be articulated by sociologists and philosophers of science in the mid-twentieth century, and it gained significant traction in the 1970s as scientists in the US were fighting to preserve their role in helping to determine which scientific projects would receive federal funding. Scientists cast review by a scientist's peers as the only legitimate method of judging the merits of a grant proposal. "Peer review," they argued, was central to properly functioning science, and if scientists were not consulted, funding decisions would be both wasteful and illegitimate. The belief in peer review's centrality endured and spread in the late twentieth century, expanding to other countries and to fields beyond the sciences, eventually resulting in headlines declaring that peer review has made a discovery "actual science."

Before I turn to my chapter outline, I want to emphasize two important things about this book. First, *In Referees We Trust?* is specifically about the rise of peer review's link to scientific legitimacy, not a full global history of the practice up to the present day. My focus is on how refereeing developed and where we got the idea that peer review is integral to science. The book will begin in Europe, where the first refereeing systems emerged, but the Cold War US turns out to be the crucial context for the development of the link between peer review and scientific legitimacy, and much of the book will therefore focus on the US. Although I do discuss the international spread of peer review, telling the full story of how peer review was adopted globally is beyond the scope of this book. That would be a very desirable topic for future scholarship, however, and I hope to see other such books and studies in the future.

Second, this book contains both discussions of actual refereeing systems and discussions of peer review discourse—what scientists *say* about peer review as a concept. As we will see, the discourse was often not a faithful reflection of how peer review worked in practice. In chapter 4, for example, we will see the physicist and philosopher John Ziman describing refereeing as the "lynchpin" of scientific credibility during an era when most scientific papers were printed without ever seeing a referee. Looking at those gaps between peer review as an ideal and peer review as actually practiced, I think, reveals the source of some major tensions surrounding peer review in modern science.

CHAPTER OUTLINE

The first chapter of *In Referees We Trust?* discusses the nineteenth-century European origins of refereeing. As it turns, out, "Who invented refereeing?" is not an easy question to answer; the general idea of scientific peers commenting on one another's work is, of course, quite old. If we define refereeing specifically as the practice of soliciting reports from one or more experts in an author's field before publishing an author's work, however, it seems clear that we can trace the origins of refereeing to nineteenth-century learned societies in Great Britain. The first experiment with referee reports was aimed at generating publishable commentaries on papers submitted to the *Philosophical Transactions of the Royal Society of London*. However, the Royal Society quickly decided that referee reports were more useful for a different purpose: providing confidential advice to the members of the society who determined which papers would be printed.

The members of the Royal Society who worked on society publications found such advice valuable because it helped create a shared sense of responsibility for editorial decisions among Fellows, and because it helped guard the Royal Society against accusations of favoritism when it came to publication in the *Transactions*. But refereeing was not seen as a guarantee of scientific truth or reliability, and its use did not spread much beyond learned societies. For-profit publications such as *Philosophical Magazine* or *Nature*, for instance, deliberately eschewed refereeing. Those periodicals had to meet inflexible publishing deadlines and often could not afford to wait to decide whether to accept or reject a submission. Instead, those publications relied on an editorial staff, with the editor-in-chief or editorial board exercising the greatest influence over the periodical's content. Notably, publications without refereeing were not seen as less trustworthy or less scientific than publications that used refereeing systems, and as of 1900, there seems to be little indication that any commentators felt refereeing could or should serve as a test of scientific legitimacy.

Chapter 2 looks at refereeing in the United States in the nineteenth and early twentieth centuries, examining why institutions in the US adopted—or did not adopt—the practice of refereeing. Learned societies in the US such as the American Physical Society generally used refereeing for the same reasons as their British counterparts: Referee reports provided social cover if the editor wished to reject a paper by a prominent figure, and learned societies also wished to avoid printing anything obviously wrong or embarrassing. US funding bodies such as the Rockefeller Foundation, on the other hand, explicitly preferred to leave decisions in the hands of funding body employees. Funding bodies did consult with scientific experts about potential awards, but such consultations were not a requirement for approving funding. It is clear that as of the 1930s, refereeing was neither universal nor considered essential at either journals or funding bodies in the US.

Chapter 3 examines the expansion of refereeing after World War II, as increasingly unmanageable workloads for journal editors led more journals in the United States to rely on external referees to make decisions. The crucial context for this shift was the massive increase in federal funding for science in the US. More scientists were being trained in PhD programs, laboratories were growing larger, and scientific journals were experiencing explosive growth in size. Unable to keep pace with the workload of reviewing the number of submissions they were receiving, many journals began relying more heavily on external referees to make judgments about scientific papers. Practical considerations were also behind the creation of refereeing systems for grants at the new National Science Foundation (NSF) and the newly expanded National Institutes of Health (NIH). The bureaucrats who set up these organizations wanted to capitalize on the positive public image of scientists in the post–World War II moment, and they also felt that it was not realistic to have government employees make decisions about the scientific merit of all proposals. It is clear from correspondence and commentary by scientists that the United States was seen as unusually reliant on refereeing. Authors trained elsewhere complained of the slowness of American journals, and a few refused to submit to journals that employed referees, arguing that refereeing kept innovative work from reaching publication.

The second half of the book will explain how refereeing—or, as it was increasingly called, "peer review"—went from being an optional process to being viewed as a prerequisite for scientific credibility. Chapter 4 will demonstrate that the idea of refereeing as something critical to scientific knowledge-making came from a source that many may consider surprising: scholars in the sociology and philosophy of science. Major philosophers and sociologists of science of the 1960s and 1970s, including Thomas Kuhn, Warren Hagstrom, and John Ziman, were widely read and cited among scientists, and their scholarship had a significant impact on how scientists described their work. Columbia sociologists Harriet Zuckerman and Robert Merton were especially responsible for shaping scientists' ideas about the purpose of refereeing. Their 1971 study of refereeing at *Physical Review*, almost certainly the single most-cited article on peer review ever written, argued that refereeing was the foundation of trust and credibility in the sciences. Chapter 4 also traces the rise of the term "peer review" in the 1960s and 1970s. The term originated in the United States and was initially used just to describe grant review processes, not journal refereeing. The gradual adoption of "peer review" as a synonym for "refereeing" is not merely a linguistic curiosity; the change in terminology established that there was a narrow range of people qualified to be referees. Only the scientific peers of a grant applicant or author, after all, could possibly perform a "peer review."

Chapters 5 and 6 both center around pivotal debates over grant review at US federal funding bodies. As we saw in chapter 3, an expansion of federal funding led to increased reliance on refereeing in the US; the contraction of that funding, ironically, ended up further cementing peer review's importance. A combination of lower Cold War tensions and increasing economic problems in the United States led legislators and members of the public to scrutinize scientific spending more closely. Tensions arose between legislators and bureaucrats, who wanted to exercise more control over how federal money was spent on scientific research, and scientists, who wanted to protect their advisory role at funding bodies like the NIH and NSF. Chapter 5 will focus on a controversy over review procedures at the National Institutes of Health. In January 1973 the Office of Management and Budget (OMB), an arm of the US executive branch charged with enacting the financial portions of the president's agenda, released a report criticizing the NIH's "peer review" method of choosing which grants to fund. The OMB report argued that the NIH was giving far too much power to the scientist-reviewers and that federal employees were the appropriate people to choose which projects received funding because those employees could be held accountable for their decisions and performance. Scientists quickly came to the defense of the NIH's system, borrowing language from the sociology and philosophy of science to argue that review by fellow scientists was the key to properly functioning science.

It was a successful defense; the NIH's peer review procedures did not change in response to the OMB report. Controversies over federal grant review continued well into the 1970s, however, and the strategy the scientific community had used to defend NIH peer review soon began to appear in other contexts. Chapter 6 focuses on a congressional hearing about grant review at the National Science Foundation. Legislators from both sides of the aisle had made public attacks on specific grants the NSF had chosen to fund, arguing that the NSF was not acting in the country's best interests and proposing that Congress should have to approve individual grants before any money was spent. NSF leaders and the US scientific community vigorously pushed back against that idea and said the NSF's peer review process was sufficient to ensure money was being spent wisely. In mounting their defense of the NSF's decision-making, scientists and NSF leaders argued that peer review was not just a good way to choose between grants but the *only* legitimate way to determine which proposals had the greatest scientific merit. The NSF controversies were a very public episode that received a great deal of press coverage, and the congressional hearings about NSF peer review both spread and helped cement the idea that peer review was central to science and the key to ensuring scientific quality.

Chapter 7 examines the aftermath of that shift, as the contraction of federal funding for science continued to be a matter of concern for scientists in the 1980s. By this time the term "peer review" was used for refereeing at both funding bodies and journals, and the idea that a journal article had to be peer reviewed to be scientifically credible shaped hiring, promotion, and tenure decisions in academic science. The process of soliciting external letters during tenure and promotion decisions was itself also increasingly called "peer review." All of this made "peer review" more fraught for young scientists trying to make careers in an increasingly competitive field. Some scientists began to worry that peer review placed minoritized scientists, and scientists at less prestigious institutions, at a disadvantage for both publication and funding. The heightened emphasis on peer review also led to an unexpected consequence: Scientists and non-scientists alike began to argue that the referee system could—and should—detect scientific fraud. However, high-profile incidents in the biomedical sciences suggested that peer review alone could not keep flawed or fraudulent results out of the scientific literature. As the end of the twentieth century drew near, there was an uneasy sense that peer review was not quite up to the tasks that had been assigned to it. In tying peer review to scientific credibility, scientists had unintentionally invested it with epistemological functions that few truly believed it could perform. But despite the complaints about it, peer review was now so deeply embedded in scientific institutions that few major changes took place.

My conclusion, written in the summer of 2025, reflects on what we can learn from the history of peer review and how that history relates to ongoing developments in US science. At the time of writing, the new US presidential administration was making sweeping cuts to many of the institutions discussed in this book. Those cuts are not just financial; the Trump administration has also paused several peer review systems at funding bodies, fired expert proposal reviewers, and suggested that they may forbid government scientists from publishing articles with independent journals. Such moves highlight the fact that peer review is not an untouchable, unalterable feature of science—and they also highlight that not all changes to current peer review systems would be positives for the scientific community. As scientists consider what they want out of peer review at a moment of intense upheaval for the US scientific community, I suggest that peer review's history points to two important questions: First, what is peer review for, and second, to whom is peer review accountable?

THE PAST AND THE FUTURE OF PEER REVIEW

Understanding how peer review developed and why it became considered the key to scientific credibility is not merely about explaining the inner workings of the scientific community. Peer review is central to how scientists present their work to stakeholders outside science, and the conception of peer review as a stamp of scientific authenticity shapes stakeholders' understanding of how science functions. The idea that peer review is supposed to bestow the status of "actual science" is also central to why many observers believe that it is currently failing in its duties. Recent scandals and studies have highlighted that many scientists have been subjected to unhelpful and personal comments from their referees. In 2015, to cite one notorious example, two female scientists writing about publishing patterns of male and female postdocs received a referee report from *PLOS ONE* saying the paper needed at least one male collaborator to prevent the analysis from "drifting too far away from empirical evidence."[18] A recent study of "unprofessional" refereeing quotes reviewer comments that include "What the authors have done is an insult to science," "This paper is, simply, manure," and a declaration that the reviewer did not read the manuscript because "the author's last name sounds Spanish."[19] On the lighter side, there are several social media accounts dedicated to lambasting "Reviewer 2" (editors usually label the most critical report as the second one) for sins ranging from not reading the paper in question to suggesting nonsensical or contradictory revisions.[20]

The cases above could be dismissed as a few bad actors mucking up a basically good system. But any search of literature on peer review will turn up dozens of articles arguing that the process is not working the way that it's supposed to on a systemic level. Critics have argued that peer review is failing at the task of detecting flawed or fraudulent research,[21] that women and underrepresented minorities are more likely to receive unfavorable referee reports than their colleagues,[22] and that current peer review procedures suppress innovative research.[23] All of those arguments speak to a particular vision of what peer review is supposed to do for science—a vision that, as this book shows, is far younger than many observers realize.

I opened this book with a story about my own experience not because I think that experience was unusual but to acknowledge something important: For academics, peer review can't help but be a personal topic. Our careers quite literally hang on the outcomes of peer review processes, and unsurprisingly, given the stakes, we tend to have complex beliefs and feelings about the institution. The referees on my own work have been largely generous, thoughtful, and constructive—the referees for this book were especially excellent—but as we saw above, many researchers have tales of referees who don't fit that description. Whenever I tell fellow scholars what I'm writing about, they nearly always ask how I think we could "fix" the peer review system—by which they generally mean weed out unprofessional reviews, ensure fairness to researchers from all kinds of backgrounds and institutions, reward risk-taking without approving sloppy or crackpot work, and preferably make sure the whole process doesn't slow down the process of applying for a grant or submitting to a journal. Unfortunately, studying the history of peer review has not given me an easy answer to that question.

This book will cover more than two centuries of history at institutions ranging from the Académie des Sciences in Paris to the US Office of Management and Budget, with a cast of characters that includes a tongue-tied Victorian physicist (George Gabriel Stokes), a publicity-seeking US senator (William Proxmire), and the exhausted 1950s editorial board of the journal *Science*. It is a work of historical scholarship that focuses on how we got to where we are, not what might be wrong in the present or where we should go in the future. But I am conscious that I am writing about the history of a process that still exists, and one that determines the career trajectory of millions of researchers. I am also aware that as I type this, many of the systems whose development I chronicle in this book are being dismantled by the current US presidential administration. My hope is that this book will help researchers in every field rethink some widely held assumptions about peer review—not with an eye toward getting rid of a process that I think has significant merit but with the goal of really considering what we want from peer review and why at a moment that seems increasingly crucial for the future of US science.

Our modern system of peer review was not handed down from on high one day during the Scientific Revolution. It developed slowly and somewhat haphazardly, has transformed significantly over time, and has been shaped by the concerns of a variety of stakeholders both within and outside of science. The more scholars consider how peer review has changed in the past, the more it becomes possible to imagine its different potential futures—both the ones we might desire and the ones we would like to avoid.

1

THE EUROPEAN ORIGINS OF THE REFEREE

A reader picking up a book on the history of refereeing might reasonably expect to learn the name of the person or people who first invented that practice. In 1971, the sociologists Harriet Zuckerman and Robert Merton suggested that it was Henry Oldenburg, the founding editor of the *Philosophical Transactions of the Royal Society*, who first consulted outside readers about manuscripts submitted for publication.[1] In this version of the story, peer review was created at the same moment as the world's first scientific journal. However, Zuckerman and Merton seem to have misread or misinterpreted Oldenburg's correspondence in reaching this conclusion.[2] In reality, Oldenburg held such close control over the journal's contents that he occasionally referred to himself as its "author." Oldenburg printed what interested him and what he thought might be of value to his readers, without a formal submission process; early issues of the *Philosophical Transactions* include not only accounts of experiments but secondhand stories about others' scientific work, discussions of recent books, and Oldenburg's own personal correspondence. More recent historical work has shown that he rarely, if ever, relied on the judgment of others to decide what to print.[3]

So if not Oldenburg, who deserves the title of "the inventor of refereeing"?

It turns out that "Who invented refereeing?" is a simple question without a straightforward answer. To even begin to investigate it, we first need a firm definition of what constitutes "refereeing." Is refereeing merely the idea that experts in a field should review each other's work? Is any solicitation of opinions on scientific or scholarly work "refereeing"? If we take that stance, it seems likely that the inventor of refereeing was an accountant in an ancient civilization who showed a mathematical solution to a fellow accountant and asked, "Does this look right?" But that tongue-in-cheek answer does not give us much useful insight into how refereeing has changed over time or how the term came to be associated with the practice of soliciting written reports on articles or grant proposals, much less how we got to our present situation in which findings need to be "peer reviewed" to be considered "real science."

To make the question of who invented refereeing historically useful, we need to define "refereeing" more narrowly. What I want to trace is not the general idea that scientists should comment on each other's work but the formal process of requiring an opinion from an expert in the author's field before an editor makes a decision on publication or a funding body decides who receives a grant. Even with that narrower definition, however, it's not easy to point to a specific moment in time and say "Aha! There, the first use of a referee!" We see learned societies begin to reach out to members for opinions on scientific work in the late eighteenth century. If we want a formal system of soliciting expert opinions prior to publication to deem a process "refereeing," however, we need to wait for the nineteenth century.

This chapter will show that the origins of the modern refereeing system can be found in Europe and that the first formalized refereeing procedures were developed at British learned society journals in the nineteenth century. The Fellows of learned societies adopted refereeing at their journals out of a desire to protect the reputation of the society and to create a sense of communal responsibility among society members for their most visible publications. Editors at for-profit commercial journals, however, did not adopt refereeing systems. Those journals had to meet uncompromising weekly or monthly deadlines and could not afford to slow a paper's path to publication if they wanted to have enough material to fill the next issue. Significantly, at no point in the nineteenth century did observers suggest that journals like *Nature* or *Philosophical Magazine*, which did not use referees, were less reliable than journals that did. Whether or not a journal used referees had little to do with the perceived quality of the papers in that journal; instead, social considerations like reputation or courtesy to a society's members were the factors that nudged journals toward using referees.

BEFORE THE REFEREE: EDITORIAL POWER AND REVIEW BY PEERS

The title of "world's first scientific journal" is generally bestowed upon two publications, both founded in 1665: the *Philosophical Transactions of the Royal Society of London* and the *Journal des Scavans*.[4] However, it would be too simplistic to say that "the scientific journal" is an institution that we have had since the seventeenth century. Seventeenth- and early eighteenth-century journals were not meant to be sites where researchers made new scientific knowledge claims. Instead, the first journals recorded events at meetings of learned societies, circulated snippets of correspondence, reviewed new books in natural philosophy (then the preferred English-language term for what we would now call "science"), and occasionally translated noteworthy foreign works into the periodical's native language. Occasionally, an author would describe a new scientific finding in a journal article, but that was by no means a requirement for either journals or authors.[5]

It's useful here to look at sample volumes of the *Philosophical Transactions* and the *Journal des Scavans* to see the differences between the journals we know today and the "journal" as first envisioned in the Scientific Revolution. Since first issues can be atypical of publications, let's look at the contents from 1668 for both journals. The third volume of the *Philosophical Transactions* was published under the title *Philosophical Transactions: Giving Some Accompt of the Present Undertakings, Studies and Labours of the Ingenious in Many Considerable Parts of the World*. The volume contained an article on making glass lenses for scientific experiments, summarizing the work of the "ingenious and industrious Francis Smethwick";[6] several book reviews, mostly of books written in Italian or Latin; extracts from correspondence about logarithms, tides, and blood transfusions; an article on a mathematical controversy happening in Italy;[7] and four pages of "Queries Concerning Vegetation, Especially the Motion of the Juyces of Vegetables, Communicated by Some Curious Persons."[8] There were accounts of experiments and observations, but those accounts by no means occupied the bulk of the volume.[9]

Similarly, the *Journal des Scavans*'s first editor, Denis de Scallo, used his new publication to collect a motley range of interesting material, from book reviews and translated papers to obituaries and reports of interesting inventions. Issues were published on Mondays and varied in length and contents; book reviews were the most common type of articles. The 1668 issues opened with a review of a new history of France;[10] the year's contents included extracts from physician's notes about new medical techniques he had observed on a trip to England,[11] a shortened version of a letter by the great Dutch physicist and mathematician Christian Huygens about a recent book review,[12] and an account of a "new discovery" about human vision.[13]

Early scientific periodicals were thus very different in form and purpose from most of the scientific journals we know today.[14] They were also very different in how they were compiled, edited, and distributed. Simply put, although some historians have followed Zuckerman and Merton in claiming that "peer review" originated in the Scientific Revolution, a closer examination of how early scientific periodicals chose their submissions reveals that systematic "peer review" did *not* exist, as either an idea or as a process, for early journals.[15] The historians Noah Moxham and Aileen Fyfe have described this period as the "prehistory" of peer review, highlighting the gaps between the general idea of having peers comment on each other's work and the formal systems of "peer review" that would develop later.[16]

Early in its history, the Royal Society licensed publication of the *Transactions*—in other words, it used a privilege granted by the British Crown to authorize publication of the periodical. The Council of the Royal Society was, in theory, responsible for reviewing the *Transactions*' contents before printing, but in practice there was little (if any) prepublication review by anyone but Oldenburg. The *Transactions* was not truly an official Royal Society periodical but instead was under both the financial and editorial control of its editor and would remain so until the mid-eighteenth century. Furthermore, Oldenburg did not sit down with a pile of submissions and decide which ones were worthy of publication in the *Transactions*. Instead, Oldenburg was constantly on the lookout for things he could print and actively requested potential material from authors. Unsolicited submissions would have been the exception rather than the rule. There is no record of any submission to the *Transactions* being rejected during the seventeenth century.[17]

In 1752, the Royal Society officially took over editorial and financial management of the *Transactions* following the death of the editor Cromwell Mortimer and some embarrassing criticisms of the papers printed in the *Transactions*. A new set of statutes outlined a formalized procedure that the society was supposed to use to approve papers for publication. Every six weeks, members of the Council of the Royal Society read abstracts of all papers communicated to the Royal Society (i.e., written by a Fellow or submitted by a Fellow on the author's behalf). The council members then voted by secret ballot to decide which papers to print. The goal of the vote was to weed out obviously unsuitable submissions rather than to limit acceptance to the best papers; the majority of communications were eventually published.[18] There was no discussion among council members, no written reports on the paper's quality, and no comments furnished to the author save for news of the council's decision—although, in the decades to follow, a secretary or president of the Royal Society might have a polite word with an author about withdrawing a flawed or potentially embarrassing paper.[19]

Across the English Channel, the French Académie des Sciences had begun experimenting with written reports on submissions in the eighteenth century. The French Académie des Sciences, founded in 1666 by King Louis XIV's famed minister Jean-Baptiste Colbert, had more official responsibilities to the Crown than the Royal Society did in England. The 1699 revised Académie statues tasked the Académie with overseeing all of French science and technology, including deciding which new technologies merited commercial protections (something like a patent) and which new writings in natural philosophy were worthy of notice.[20] In 1717, King Louis XV specifically granted the Académie the right to publish scientific texts "after examining the works and judging that they are worthy of publication," making review of books and papers prior to publication an explicit job of the institution.[21] In the eighteenth century, the task of deciding what the Académie would publish was given to the Comité de Librairie, a ten-person committee of Académie insiders.[22] Only papers that had not been previously published elsewhere could be considered for publication. As part of its decision-making process, the Comité would ask two or three members of the Académie to write reports closely examining the paper's claims and making a recommendation about publication.[23] The authors of those reports were called *commissaires*. Authors of papers were expected to make revisions in response to the comments of the *commissaires*, and the Comité de Librairie explicitly saw itself as responsible for confirming the quality and accuracy of the papers published by the Académie.[24]

So far this will sound quite familiar to most modern scientists; as the historian James McClellan argues, the eighteenth-century Académie des Sciences seems to be an "early and weighty episode" in the history of refereeing.[25] However, it also seems to have been a somewhat isolated episode. The Académie's system did not catch on with eighteenth-century contemporaries. Other institutions such as the Société Royale de Médecine, the Académie Royale de Chirurgie, and Britain's Royal Society explicitly said that they did *not* guarantee the accuracy of papers or books published under their name. Responsibility for quality and accuracy lay at the author's feet, not with the learned society.

Furthermore, the Académie's eighteenth-century system does not seem to have directly led to the systems of external refereeing that would develop in the nineteenth century. In 1793 the Revolutionary government disbanded the government-sponsored Académie. Its successor, the Académie des Sciences of the Institut de France, did not continue the Comité des Librairies that had overseen the thorough review procedures of the eighteenth century. Indeed, the new Academie's most influential nineteenth-century periodical, the *Comptes rendus hebdomadaires des seances*, deliberately emphasized fast publication rather than comprehensive review. Académie outsiders could still submit their work to the Académie and receive a *rapport* about its contents—but a *rapport* was significantly different from both the eighteenth-century reports written by *commissaires* and from modern referee reports. A nineteenth-century author's goal in sending a paper or book to the Académie was often not to see it printed by that organization but to receive the report itself. Many authors used Académie reports as prefaces so that they could boast of the

Académie's praise for their work, making submission to the Académie something like the nineteenth-century equivalent of asking for a book blurb. An author who submitted their work to the Académie was *not*, notably, submitting it for the judgment of his peers. Instead, he was submitting it to his *superiors* in the hierarchy of French science in the hope that they would find it worthy of praise.[26]

FROM PUBLIC PRINTING TO PRIVATE DOCUMENTS

Alex Csiszar's *The Scientific Journal: Authorship and the Politics of Knowledge* convincingly demonstrates that the modern scientific journal was a creation of the nineteenth century. Although periodicals devoted to science existed before then, it was not until the nineteenth century that papers in journals became "the chief avenue by which to make the results of [scientific] research into public knowledge claims."[27] Csiszar also traces the origins of the "scientific referee" to nineteenth-century Britain, where "men of science" (their preferred term for themselves) were engaged in a struggle for social and intellectual legitimacy in their class-conscious society.[28] Learned societies, including the Geological Society, the Astronomical Society, and the Royal Society itself began to experience pressure to restrict admission and publication to qualified researchers, rather than welcoming dilletantes whose main qualification was the ability to pay dues. That meant these societies needed a way of distinguishing between the qualified and the unqualified.

Some British critics looked to the French system of official reports by Académie *rapporteurs* with envy, believing that rigorous public evaluation of new scientific claims would enhance both the quality of British research and social and intellectual respect for men of science. Several British learned societies began experimenting with new forms of prepublication vetting in the early decades of the nineteenth century. The Geological Society of Great Britain seems to have been the first to use the term "referee" in conjunction with prepublication review. In 1817, the Geological Society's first president, George Greenough, decreed that all papers that the council deemed potentially of interest would be sent to a "referee," an expert in the paper's field who could recommend changes that might improve the paper. In practice, however, the reports were short and usually contained only a recommendation on whether the paper should be published.[29]

Around the same time, the Astronomical Society of London established committees that made decisions on all papers submitted for publication, rather than selecting individual commentators for each article.[30] Similarly, by 1830 the Royal Society was using a Committee on Publications to decide what would be printed in its *Transactions*. However, the members had not been entirely content with that system. In an 1830 pamphlet titled *Science Without a Head*, Augustus Granville complained that the Committee on Papers included Fellows who had "not the smallest pretension to any knowledge whatever of the subject."[31] Interestingly, Granville claimed that the committee was supposed to share every paper with "some competent member," but both Granville's pamphlet and the records of the Royal Society indicate that this was not done systematically and was more of an aspiration than an actual policy.[32]

In February 1831, the Fellowship of the Royal Society of London unveiled a new periodical: the *Proceedings of the Royal Society of London* (henceforth *Proceedings*).[33] The reason for the society's first new journal in nearly two hundred years was simple. Fellows of the society, other men of science, and interested members of the public were increasingly unwilling to wait for the slow-publishing *Transactions* for news of what had happened at the most recent Royal Society meeting, and the Royal Society's leadership was unhappy that commercial publications had begun to fill the gap by printing news of the Royal Society's meetings without the society's formal approval. *Proceedings* was intended, in the words of the newly elected Council of the Royal Society, to circulate "the Abstracts of Papers read at the Society's Meetings and entered upon the Minutes of their proceedings . . . for the use of the Fellows."[34] The new journal was an immediate hit, with many requests for copies and reprints.

Like most new periodicals, the *Proceedings* needed to find content, and the Cambridge polymath William Whewell came up with a plan to provide some.[35] In a March 1831 letter to his protégé John William Lubbock, Whewell suggested that the Royal Society should solicit and print commentaries on the papers submitted to the *Transactions*.[36] Those commentaries could then be printed in *Proceedings*, giving readers even more reasons to pick up new issues of the society's new publication. Whewell told Lubbock that he had been inspired by the occasional publication of Académie *rapports*.

The other Fellows evidently liked Whewell's suggestion. They agreed to solicit reports on a paper by a young astronomer named George Airy (later famous as the Astronomer Royal who made Greenwich the location of the prime meridian). Lubbock and Whewell themselves agreed to write reports on Airy's paper "On an Inequality of Long Period in the Motions of the Earth and Venus." However, as Alex Csiszar recounts, Whewell and Lubbock came to opposite conclusions on the paper. Whewell admired the paper's mathematics and thought them solid; Lubbock, on the other hand, disliked Airy's introduction of an additional term into a Lagrangian theorem. The addition, Lubbock complained, "destroys entirely their [the equations'] analytic beauty."[37]

The disagreement was a problem because the plan had *not* been for Whewell and Lubbock to each write a report. They were supposed to coauthor a single document that both of them would sign. Lubbock apparently paid a visit to Airy to see if Airy might edit the manuscript to bring him and Whewell into agreement, but Airy was unwilling. "I must say that I do not particularly admire being put on my trial in this manner," he told Whewell in a personal letter. "There the paper is and I am willing to let my credit rest on it."[38] Eventually, Whewell agreed to some changes and Lubbock agreed to sign his name. The joint report was presented at an 1832 meeting of the Royal Society and, eventually, printed in an issue of *Proceedings of the Royal Society* alongside a similar report about a recent paper by Michael Faraday.[39]

The Royal Society's new president, Prince Augustus Frederick, the Duke of Sussex, hoped that this new practice of soliciting and printing coauthored reports on papers would continue. Other members of the society, however, were apparently not so enamored. The Royal Society had a long-standing tradition of what Fyfe and her collaborators call "gentlemanly politeness;" if the society authorized the printing of a report that criticized a Fellow's paper, it would go against every unspoken rule about how the fellowship should conduct themselves in public.[40] Even Sussex agreed that only positive reports could ever reach print. Sussex also proposed that only members of the Royal Society's Council—its senior leadership, twenty-one Fellows elected by the membership—should coauthor reports. That way, the authors whose papers were being commented upon would know that the commentary came from the Royal Society's most eminent minds. But the proposal to have council members write the reports quickly encountered a problem: The members of the council found it burdensome to add referee reports to their workload, especially if they would have to come to agreement with another member of the council on its contents.

And so the Royal Society quickly pivoted to a new system. Instead of requiring coauthored public reports written by members of council, the Royal Society leadership decided to seek out confidential reports from a broader pool of Fellows. In 1833, the Royal Society established committees of Fellows in various scientific disciplines and tasked those committees with awarding Royal Medals in their fields; by 1838, the committees were also charged with evaluating submissions to the *Transactions* in their fields. Committee members reviewed the original manuscripts of submitted papers and were allowed to send those manuscripts to experts who were not committee members. The Royal Society's two secretaries—one in the biological sciences and one for the physical sciences—coordinated the practical aspects of the refereeing system, including sending papers to the referees, collecting referee reports, and communicating with the scientific committees and the Committee of Papers about the referees' recommendations.[41] As Whewell put it in an 1836 letter, the new system of anonymous reports "gives us a useful and indeed indisputable means of excluding from the Transactions papers which for matter or form, ought not to appear there."[42]

Even with this system in place, however, we should not make the mistake of assuming that the Royal Society's refereeing procedures looked like the ones we are accustomed to seeing at modern journals. First, to be considered for publication at all, a paper had to be read aloud at a meeting of the Royal Society by a Fellow. Researchers outside the fellowship would have to ask a Fellow to read their work for them (a system also in place at the Académie des Sciences). Some referees were reluctant to ask for too many revisions or corrections for fear of making the published version too different from the paper read aloud at the meeting. The scientific committees also tended to rely on a small number of highly productive referees. Fyfe and her collaborators found that half of the surviving referee reports from the 1830s were written by the same fifteen Fellows.[43] (The most productive? George Airy, whose paper was the subject of that first coauthored report by Whewell and Lubbock.) In 1839, the decision about which papers to publish in the *Transactions* acquired new significance when the Royal Society reformed its method of choosing new Fellows. The new guidelines required a list of a potential Fellow's publications, and having a paper in the *Transactions* was considered an especially noteworthy qualification for fellowship.[44] Therefore, having a paper chosen for the *Transactions* was not simply a way to publicize one's recent work—it was a way for researchers to establish themselves as fully credible men of science and potentially gain entry to Britain's most prestigious scientific society.[45]

The Royal Society leadership was now under increased pressure to persuade the members that papers were approved for publication in the *Transactions* on the basis of solid scientific work and that authors were being treated fairly. That pressure grew after a public scandal surrounding the 1845 Royal Medal in Physiology. The awardee, Thomas Beck, was the protégé of council member William Sharpey, and his paper was given the Royal Medal before appearing in the *Transactions*—in fact, Beck's paper was only approved for publication on the day it was awarded the Royal Medal. Many Fellows were outraged (especially those who felt they had stronger claim to the Royal Medal), and the widely read medical journal *The Lancet* castigated the Royal Society for its conduct. The council ultimately voided the 1845 medal in 1847; however, the episode intensified both internal and external attention on how papers reached the pages of the *Transactions*.[46] It seems likely that this controversy was, at least in part, responsible for a change to the society's refereeing system: Scientific committees would no longer be responsible for arranging refereeing. Instead, that task would fall more heavily on the secretaries.

GEORGE GABRIEL STOKES AND REFEREEING IN THE MID-CENTURY

The person who most shaped refereeing at the mid-century Royal Society was the physicist George Gabriel Stokes (figure 1.1), who served as the physical sciences secretary from 1854 to 1885. The Irish-born Stokes

1.1 George Gabriel Stokes (1819–1903), who served as the physical sciences secretary of the Royal Society of London from 1854 to 1885. (Courtesy of Pembroke College, Cambridge; CC BY-SA 2.0 DEED.)

matriculated at Pembroke College, Cambridge, in 1837 and quickly climbed the ladder to Britain's scientific elite by becoming First Wrangler, the student who achieved the best score on the demanding mathematics tripos examination. Pembroke quickly offered him a position, and in 1849 Stokes was named to Cambridge's Lucasian Professorship of Mathematics, the post once held by Isaac Newton. Two years later Stokes was elected to the Royal Society of London.

Stokes accepted the role of secretary in 1854, likely because it came with a salary of £105 per year that would supplement Stokes's Cambridge income.[47] He initially believed the job would be little work, but it soon became one of his life's major undertakings.[48] Although Stokes was nominally the physical sciences secretary, his work as an editor and administrator was not confined to the physical sciences. In 1884 he would tell the eminent biologist Richard Owen, in no uncertain terms, "I am editor of the Transactions."[49] That line is notable because Stokes did not often make such stark claims about himself. Stokes was famously quiet and modest; during his thirty years as secretary, the Royal Society Minutes record Stokes speaking only thirteen times.[50] Nevertheless, communication was central to Stokes's scientific life. David Wilson has characterized Stokes as a "Victorian correspondent," someone whose very career lay in the receiving and writing of letters.[51]

As a secretary of the Royal Society, Stokes exchanged letters not only with the other Royal Society leaders and with the referees he asked to write papers but with the authors of the papers themselves. Stokes's collegial, collaborative approach nicely illustrates the way in which, as Aileen Fyfe, Flaminio Squazzoni, Didier Torny, and Pierpaolo Dondio put it, refereeing helped establish "collective responsibility for editorial decisions" among members of the Royal Society.[52] Elsewhere, Fyfe and Anna Gielas have called this sort of arrangement "distributed editorship," a system that transformed publication in a periodical like the *Transactions* into "something that represented a kind of community approval."[53] The society's reputation rested in part on the reputation of its publications, and therefore, Fellows had "a moral obligation to contribute to the editorial process."[54]

Significantly, refereeing was far more likely for papers submitted to the Royal Society's crown jewel, the *Philosophical Transactions*, than for papers submitted to the relatively young *Proceedings*. Between 1865 and 1900, over 80 percent of papers published in *Proceedings* were not sent to any referees. In contrast, over 80 percent of the papers published in the *Transactions* during that period were sent to at least one referee—a figure that indicates how important the *Transactions* was to both the society's reputation and to the authors who hoped to publish their work in its pages.[55]

Referee reports at the nineteenth-century Royal Society ranged from multi-page critiques of minor experimental details to brief missives informing Stokes that a paper was acceptable and should be published. Stokes also took on a great deal of the physics refereeing himself and was the most frequent author of *Transactions* referee reports.[56] The referee reports were considered confidential documents for the internal use of the Royal Society, and the author was not given a copy of the referees' remarks. However, that does not mean that authors received no feedback at all. Stokes used personal correspondence to communicate suggestions to authors and to encourage them to correct specific issues before publication.[57] The Cambridge physiologist Michael Foster, who became the biological sciences secretary in 1881 and worked alongside Stokes during the final years of Stokes's secretary term, praised Stokes's gift for turning "crude, imperfect" submissions into "something worthy of being made known."[58]

To give us a sense of how Stokes approached his job, we can look at the correspondence between Stokes and his fellow Irishman, the physicist John Tyndall. Like Stokes, Tyndall focused his research on the physical sciences, and the two often exchanged letters about interesting observations or vexing physics questions.[59] Although they were not part of each other's closest inner circles, Tyndall and Stokes were friendly—by the 1860s they began their letters with the salutations "My dear Stokes" and "My dear Tyndall"—and the tone of their correspondence is familiar and collegial. Given that connection, we might reasonably wonder whether their correspondence is typical of Stokes's approach to refereeing. However, the archival record makes it clear that Stokes communicated with many *Transactions* authors in the same courteous way, even when he was not comfortable using a socially informal salutation.[60]

When Stokes refereed Tyndall's papers, he would send Tyndall a personal letter filled with comments and suggestions. Stokes went line by line through Tyndall's submissions, questioning word choices and pointing out places where he felt Tyndall had not made his case sufficiently. For example, on May 7, 1861, Stokes sent Tyndall a twenty-four-page letter about Tyndall's paper "On the Absorption and Radiation of Heat by Gases and Vapours, and on the Physical Connexion of Radiation, Absorption, and Conduction."[61] The letter contained a numbered list of comments about Tyndall's work along with suggestions for improving perceived weaknesses in the analysis. In the final version of the paper, it seems that Tyndall took most—though not all—of Stokes's advice.[62] For instance, Stokes's first comment was about Tyndall's discussion of a similar set of experiments undertaken by a Dr. Franz. Stokes expressed some concerns about Tyndall's discussion of Franz's work, writing "It does not seem to me that there is any real discrepancy between Franz and yourself at least nothing more than is fairly attributable to the difference in the sources of heat and similar differences in the mode of conducting the experiments." In the published paper, Tyndall wrote, "The differences between Dr. FRANZ and myself admit, perhaps, of the following explanation. His source of heat was an argand lamp, and the ends of his experimental tube were stopped with plates of glass"[63]—a sentence with strong echoes of Stokes's observation.

Stokes commented more extensively on Tyndall's discussion of the relationship between radiation and conduction. "Now in the first place is it a physical law that good conductors are bad radiators and bad conductors good radiators? I confess I doubt it," he wrote, moving on to question Tyndall's apparent assumption that

> a given molecule loses the main part of its motion in the course of one or two vibrations. For you suppose that in the case of a good radiator a molecule does not lose much motion by communication to the adjacent molecules because it has already lost it by communication to the ether, which acts as a buffer. . . . But if the fraction of the vis viva of a molecule which is lost in a single vibration is very small, so that a number of successive vibrations are nearly equal, it seems to me that the loss by communication to the ether and the loss by communication to other molecules must take place sensibly independently of each other.

In the final paper, it appears that Tyndall cut down significantly on his theoretical discussion of radiation and conduction. Near the end of the paper, Tyndall asked,

Why should good conductors be, in general, bad radiators, and bad conductors good radiators? These, and other questions, referring to facts more or less established, have still to receive their complete answers. It is less with a hope of furnishing them than of shadowing forth the possibility of uniting these various efforts by a common bond, that I submit the following reflections to the notice of the Royal Society.[64]

He answered Stokes's critique by casting his paper as inspiration for further investigations, rather than as an answer to the puzzle of the relationship between radiation and conduction.

Comparing the Tyndall-Stokes correspondence with the referee reports Stokes filed with the Royal Society yields further insight into Stokes's work with the *Transactions*. A letter from Stokes to his closest friend, the famed physicist William Thomson, reveals that Stokes did not write a full referee report for the Royal Society about Tyndall's 1861 paper:

I did not write a report on Tyndall's paper but merely recommended it, stating however my dissent from the author's view as to the relation between conduction and radiation, but saying that if the author wished it retained he had done such good work that he had, I thought, a right to keep it. There were other points of less importance about which I said I would write to the author, and have already done so.[65]

This letter indicates that Stokes saved significant criticism of "On the Absorption and Radiation of Heat" for a personal letter rather than a full referee report. (Thomson also refereed the paper, but unfortunately his report has not survived.[66]) Similarly, in 1863 Stokes refereed Tyndall's paper "On the Relation of Radiant Heat to Aqueous Vapour." His official report for the Royal Society was quite brief and offered only a summary and endorsement of Tyndall's paper.[67] Once again, he saved his comments for a personal letter to Tyndall, although in this case he made only two fairly minor suggestions, one about wording, the other about a citation.[68]

In 1874, in contrast, Stokes did submit a report on Tyndall's paper "On the Atmosphere as a Vehicle of Sound."[69] Stokes and Robert B. Clifton, a physicist at Oxford, were the paper's referees. Stokes's report recommended that Tyndall's paper be printed, but he also registered a significant objection to Tyndall's conclusion that differences in humidity rather than differences in temperature changed the distance over which sound could travel through air. Stokes analyzed Tyndall's experimental setup carefully and used a number of physical equations to illustrate why he thought that temperature, not humidity, was the more important factor. He concluded:

I think therefore that Dr Tyndall is not justified by the evidence he has [amassed] in assuming so confidently as he has done that difference of humidity rather than temperature is the cause . . .

The value of the paper is little affected by the decision of this portion one way or other, but I think it important that a result should not be confidently asserted in the pages of the Phil. Trans. which rests on evidence quite insufficient to establish it.[70]

Notably, although Stokes was concerned about printing a questionable conclusion in the *Transactions*, he did not feel that his objections were grounds for rejecting the paper—it was still "valuable" even if Tyndall's conclusions were incorrect. And in fact, Tyndall's paper was accepted for publication, even in light of Stokes's reservations.

In a letter to Tyndall dated March 19, 1874, Stokes enclosed both a form letter informing Tyndall that his paper had been accepted and a copy of Stokes's referee report. He also related one of Clifton's comments from the second report filed with the Royal Society. Stokes did not mention Clifton by name, further indicating that although referees appeared to know who had written the papers under consideration, authors would not know the names of the referees unless (like Stokes) they chose to make themselves known.[71]

Once again, Tyndall changed his paper after receiving Stokes's suggestions. The final version of "On the Atmosphere as a Vehicle of Sound" placed much more emphasis on the importance of difference of temperature than on differences of atmospheric conditions such as fumes, fog, or humidity—except when measuring sound traveling over water, a case that Stokes had conceded might be influenced by the presence of vapor.[72] Tyndall also included a sixteen-page appendix collecting many detailed observations of the atmospheric conditions on various days and lengthy descriptions of observed phenomena.

Stokes's letters and his referee reports about Tyndall's papers show that there was a significant difference between Stokes's use of official reports and his use of personal correspondence. When Stokes felt

Tyndall's article was basically acceptable, as he did in 1861, he relied on personal correspondence to relate suggestions to Tyndall and either filed no referee report or only a short summary of the paper with the Royal Society. Stokes's objections to Tyndall's 1874 paper on sound, however, were more serious, and in this case, he did file a report with the Royal Society registering his criticism. However, he still used correspondence to ensure that Tyndall received a copy of that report and to relate a comment from one of the other referees that he felt was important. All of this indicates that for Stokes, referee reports and personal correspondence served two very different purposes in his editorial work. The referee report was a document for the internal use of the Royal Society; it was meant to ensure that the papers in the *Philosophical Transactions* were scientifically respectable and that their selection had not been the result of favoritism. The report was not necessarily meant for the eyes of the paper's author. Even if a paper met the criterion of scientific quality, however, Stokes often felt that it could be improved and used personal correspondence to offer specific criticisms of an accepted paper.

As we can see, while Tyndall did not incorporate all of Stokes's suggestions exactly, he frequently changed his papers in response to Stokes's feedback. The correspondence between the two men indicates that Tyndall was not merely bowing to pressure in order to get his papers into print. Stokes's 1861 letter opened with the statement "Your paper was ordered to be printed last Thursday," and he indicated that the plate for Tyndall's figure was already being prepared; it was clear that any further corrections to the text of the paper would be at Tyndall's discretion. Similarly, the 1874 paper had already been accepted when Stokes sent a copy of his referee report to Tyndall.

It is possible that Tyndall implemented the suggestions to ingratiate himself with Stokes, but it is clear from the rest of their correspondence that Tyndall respected Stokes's scientific judgment. Stokes was obviously a thorough reviewer, and it seems that Tyndall, in turn, took Stokes's suggestions seriously. After Tyndall received Stokes's May 7 comments on "On the Absorption and Radiation of Heat" in 1861, several more letters discussing and clarifying Stokes's suggestions passed between the two physicists.[73] Tyndall defended his conclusions in their correspondence, but the published paper still eliminated most of Tyndall's previous claims about the relationship between conduction and radiation.

Similarly, in June 1874 Tyndall and Stokes engaged in a lively discussion of Tyndall's proposed revisions to "On the Atmosphere as a Vehicle of Sound." Tyndall's letter prompting the exchange has not survived, but Stokes wrote a reply on June 11, 1874. His letter suggested that Tyndall had taken some exception to his comments in the referee report, and that Tyndall may have gone so far as to quote or reference Stokes's objections in the revised paper. In a seventeen-page letter, Stokes mounted a spirited defense:

> My report was sent you that you might adopt such suggestions therein contained as you might agree with. It is not desirable that you should enter into controversy against arguments that you may not see reason to agree with in an unpublished paper, so that the public are not in a position to weigh fully what is to be said on both sides.[74]

By Stokes's standards, the letter was quite pointed. In one paragraph, Stokes told Tyndall that he had "misconstrued the sentence quoted . . . You are therefore arguing against a $\sqrt{-1}$ [i.e. an imaginary] opponent." In another, referring to Tyndall quoting Sir John Herschel as an authority on atmospheric conditions, Stokes wrote, "It is useless to shelter oneself under the authority of a great name in a matter in which, once a doubt is started, there is a direct appeal to simple arithmetic."[75]

Tyndall thanked Stokes for his remarks in a letter dated June 17; however, he did remark that the tone was not quite what he expected from Stokes.

> I am very much obliged to you for the trouble you have taken both in the Report, & in your last letter. Let me frankly say however that there is an expression in your paragraph "ß" that I would rather see not there. However I dare say no sharpness on your part was intended.[76]

The comment Tyndall objected to was Stokes's remark about "shelter[ing] oneself under the authority of a great name." In a reply dated June 21, Stokes asked Tyndall to consider the phrase reworded and replace "shelter oneself under" with "invoke."[77]

The exchanges between the two men do not suggest an unapproachable, dictatorial editor, but rather a colleague whose scientific opinions

Tyndall respected. Tyndall felt confident in clarifying some points, asking for clarification on others, and even pointing out an occasion when he thought Stokes had been rather too harsh. Another letter makes Tyndall's views on the value of Stokes's judgment even more explicit. Although the Stokes letter that prompted this reply appears to be missing, Tyndall's undated letter implies that Stokes had expressed concern that Tyndall might feel himself unjustly pressured to change a paper according to Stokes's suggestions. Tyndall assured Stokes that this was not the case, writing that he "had no thought of 'coercion'" but believed that "any point on which you have thought, and regarding which you have arrived at an opinion opposite to mine, demands from me very careful consideration before committing myself to print upon the subject." Tyndall then said he preferred to delete the passage Stokes had questioned.[78]

REFEREEING AS THE EXCEPTION, NOT THE RULE

Stokes's tenure as secretary coincided with a significant period of transition for specialist scientific periodicals, the journals and magazines written by and for researchers in the sciences.[79] The number of scientific periodicals of all kinds experienced an enormous increase over the course of the nineteenth century. In Britain, for example, historians estimate that there were at least ten times as many periodicals dedicated to science in 1900 as there had been in 1800.[80] This development was not greeted with tremendous enthusiasm by members of learned scientific societies. Many of those new publications were written by science journalists, a relatively new class of writer who may or may not have had significant training in the sciences. Furthermore, some for-profit journals began printing accounts of learned society meetings or even preprints of papers that would soon appear in a learned society journal—a practice that most learned society leaders found annoying at best and gentlemanly misconduct at worst.[81]

Despite the discomfort with the new journalistic periodicals, scientific researchers increasingly published their work in the form of articles. In the mid-nineteenth century, it was common for men of science to publish their most significant research as monographs or memoirs; one famous example here is that of Charles Darwin's *On the Origin of Species*, published in 1859. By 1900, however, men of science (or, as they were increasingly called, "scientists") were building their careers and reputations by publishing in specialist journals. To be a scientist was to be an author of scientific articles. The increasing emphasis on authorship of papers over the course of the nineteenth century was one of the reasons that learned societies like the Royal Society faced increasing pressure to show that submissions were chosen fairly and based on merit. When someone's career depended on publication, the stakes for being accepted or rejected from a journal became much higher.

It is important to emphasize, however, that learned societies' decision to use refereeing did not usher in an era of universal refereeing at scientific journals. Well into the twentieth century, in fact, referee systems were the exception, not the norm. As noted above, although the Royal Society consulted referees about the vast majority of the papers in the *Transactions*, papers published in the newer *Proceedings* only rarely went to referees before 1900. The Royal Society's small but growing program of grants for scientific research also did not use an external refereeing system or solicit reports on grant applications, as recent research by Danielle Farrier has shown. Instead, formally appointed committees composed mostly of Fellows chose which applicants would receive money, and there seems to have been little to no attempt to avoid what we would now call conflicts of interest. Committee members frequently reviewed the applications of friends and protégés, and until the 1880s, they could even apply for funds themselves.[82]

A useful example of a non-refereed journal is *Nature*, a London-based scientific magazine founded in 1869 by the astronomer Norman Lockyer. Lockyer, who was *Nature*'s editor-in-chief for its first fifty years of publication, founded the journal in the hopes that it would become a venue where British men of science could extoll the achievements of British science to an audience of elite, educated laymen. However, by the 1880s, *Nature*'s contributors had remade the journal into a periodical by and for scientific specialists. *Nature*'s weekly publication schedule proved especially useful to contributors as a means of rapid publication for correspondence, debates, and new scientific findings.[83]

Nature was different in significant ways from any of the scientific periodicals discussed so far in this chapter. Periodicals affiliated with learned societies published slowly and on an irregular schedule; they went to print when there was enough material to warrant a new issue. Like other commercial scientific magazines, such as the *Philosophical Magazine* and *Chemical News* (its most direct inspiration), however, *Nature* published regularly—in *Nature*'s case, on an unforgiving weekly schedule. Furthermore, in the nineteenth century, learned societies did not expect to make money from journal publication—journals were regarded as a service and an expense, not a potential source of income. Stokes did not have to worry very much about finances, printing, distribution, or any of the myriad practical business concerns associated with publishing a periodical. In contrast, Lockyer was responsible for overseeing every aspect of *Nature*'s production and was under intense pressure to turn a profit for *Nature*'s publisher, the London publishing house Macmillan and Company.[84]

As noted earlier, editorial responsibilities at *Philosophical Transactions* were distributed among several people, including secretaries, referees, and members of the Committee on Papers. Stokes was highly influential, but his sphere of responsibility was carefully defined, and his power over *Philosophical Transactions* paled next to the near-total control exercised by a sole editor like Lockyer. Although *Nature*'s publishing archives have unfortunately not survived, it is clear from personal correspondence between Lockyer and his contributors that *Nature*, like other commercial journals, did not employ referees. Lockyer himself—occasionally with the help of an assistant—made the decisions about what *Nature* would publish and considered himself perfectly qualified to do so whether the subject was astronomy (his own field) or geology or botany.

Lockyer's approach was not the result of ignorance about referee systems. As a Fellow of the Royal Society, Lockyer was familiar with the procedures used at the *Philosophical Transactions*. Surviving letters to Lockyer from Stokes and Stokes's successor, Joseph Larmor, demonstrate that Lockyer received comments and suggestions for revisions on papers he submitted to the Royal Society.[85] So he knew such systems existed elsewhere but deemed them inappropriate for his own commercial, for-profit, fast-publishing journal. In the first decades of the journal, *Nature* was often struggling to fill its pages with suitable content—weeklies were content-hungry, and there was rarely, if ever, a significant backlog of submissions. To meet *Nature*'s weekly schedule, Lockyer (and later, his sub-editors) often had to make snap decisions about which submissions to print. *Nature*'s articles were not evaluated by anyone outside the editorial staff because, simply put, there was not enough time to solicit outside opinions and still get a magazine printed every week.

By the early twentieth century, scientists who submitted their work to learned society journals in the United Kingdom knew to expect that their work would be refereed by that society's members. Imogen Clarke has shown that Royal Society publications such as the *Philosophical Transactions* and the *Proceedings* continued to rely on a small number of frequent referees whose opinions had tremendous power to shape the contents of those journals. For example, in 1920, the Royal Society referees John William Nicholson and Charles Galton Darwin recommended rejecting a paper by the physicist Richard Hargreaves that they considered too old-fashioned. Hargreaves was not a Fellow and thus could not submit a paper to a Royal Society journal himself; the prominent Fellow Joseph Larmor, a former physical sciences secretary, endorsed and submitted Hargreaves's work for him. Despite Larmor's status, Darwin and Nicholson successfully argued that Hargreaves's paper should not appear in the society's journals because it embraced classical mechanics rather than the new quantum mechanics—a decision that deeply upset Larmor.[86] Under the guidance of quantum believers like Nicholson and Darwin, the Royal Society journals became institutions that supported the ongoing revolution in mechanics. Physicists who did not embrace the new physics of the early twentieth century, however, increasingly felt themselves marginalized and excluded by the society's referees.

Although scientists had grown accustomed to refereeing at learned society periodicals, that expectation did not transfer to other types of journals. British commercial publications like *Philosophical Magazine* and *Nature* continued their practice of relying on editors or a small editorial board to make decisions about the merits of a paper. Journals that did not use referees did so as a deliberate choice, not out of ignorance or unfamiliarity with the practice of refereeing. The *Philosophical Magazine* editor Oliver Lodge, for example, strongly preferred a more flexible system. In response to a 1921 letter criticizing the management of his journal as haphazard and unrigorous, Lodge declared that "it is possible to over-organise things into lifelessness."[87] *Philosophical Magazine* became something of a haven for physicists who were skeptical of relativity or quantum mechanics. Lodge himself was a classically aligned physicist and had strong professional and personal connections to many of the prominent quantum and relativity skeptics.[88] Maintaining a system of strong editorial control gave Lodge the scope to publish papers that were outside the modern mainstream of physics but that Lodge felt were valuable contributions to science.

In the case of *Nature*, introducing more extensive refereeing in the early twentieth century would have compromised one of the journal's most desirable features: its speed of publication. In fast-moving fields like Mendelian genetics and radioactivity research, *Nature* became the publication of choice because its Letters to the Editor section could print a submission within a week of receiving it. The absence of refereeing, in other words, could be a positive quality for a journal when researchers were worried about establishing their scientific priority claims.[89] Attracting papers on recent work in exciting fields of research also made *Nature* stand out from other scientific periodicals—and helped *Nature*'s sales and subscription figures, a constant topic of concern for *Nature*'s publisher Macmillan and Company.[90]

CONCLUSION: AT THE FIN-DE-SIECLE

By World War I, refereeing was well established as an expected part of publication at many British learned societies. However, the practice was far less common in other nations. Even in France, the home of the *rapport*, those *rapports* were declining in importance as referees became more visible in the UK.[91] The relatively young scientific community in the US had a handful of journals, but they largely followed the model of having a powerful editor make most of the editorial decisions. Furthermore, as we have seen, a significant number of periodicals in the UK deliberately eschewed refereeing because it might have interfered with other top priorities such as fast publication and profitability.

So referees were a creation of the nineteenth century, but they had not yet acquired the significance that later generations of scientists would bestow upon them. Scientific papers and books that were not refereed were not considered unreliable or less valid than refereed ones; something could be "real science" without ever passing through the hands of a referee. A referee system was an optional part of scientific publishing, and referees were described as figures meant to protect the honor of a specific learned society or journal, not as people performing an essential service for the scientific community as a whole.

Finally, it is worth noting that opinions of refereeing were not entirely positive, even in the first years of the practice. It took less than a decade for complaints about referees and their anonymity to begin circulating among Fellows of the Royal Society. In 1845, an anonymous article about the Royal Society in *Wade's London Review* complained that authors of papers never learned the names of the people who passed judgment on their work, nor were there "any accessible means of learning what the report itself may be!" A referee, said the anonymous writer, "may be a man of integrity" or he might be a rival of the paper's author, "full of envy, hatred, malice, and all uncharitableness," who joyfully took the opportunity to savage his adversary's work and deny it publication.[92] In March 1852, the young biologist T. H. Huxley told his sister Eliza that he feared his new paper would be refereed by a nemesis, whom he sarcastically called "my particular friend," and said that if it were, "then it will not be published."[93] In 1902 the astronomer Henry Armstrong, a vice president of the Royal Society, even proposed *replacing* referees with a dedicated editor who could make publication decisions without imposing an undue reviewing burden on the Royal Society's Fellows.[94] It was hardly a foregone conclusion, in other words, that the referee would become a figure associated with scientific reliability—and we will have to wait until much later in the twentieth century to understand where that association came from.

2

ACROSS THE ATLANTIC

REFEREEING COMES TO THE EARLY TWENTIETH-CENTURY US

In 1936, the eminent physicist Albert Einstein and his coauthor Nathan Rosen submitted a paper about gravitational waves to the American physics journal *Physical Review*, the flagship publication of the American Physical Society (APS). Einstein likely anticipated a smooth path to publication despite the paper's controversial claim that gravitational waves did not exist. After all, Einstein's previous submission to *Physical Review*, a provocative anti-quantum paper, had been quickly approved by *Physical Review* editor John Torrence Tate Sr.[1]

But to Einstein's surprise, six weeks after mailing the gravitational waves paper to *Physical Review*, he received a letter from Tate containing not a timeline for publication but a list of critical comments from an anonymous physicist. Einstein's reply betrayed more than a hint of annoyance.

> Dear Sir,
>
> We (Mr. Rosen and I) had sent you our manuscript for publication and had not authorized you to show it to specialists before it is printed. I see no reason to address the—in any case erroneous—comments of your anonymous expert. On the basis of this incident I prefer to publish the paper elsewhere.[2]

Modern readers might see Einstein's reaction as arrogant or assume that he was trying to use his fame and status to demand special treatment. But Einstein had been in the US for less than three years when he and Rosen sent that paper to *Physical Review*. Like most scientists from continental Europe, he was accustomed to receiving signed comments from a journal's editor, not anonymous reports. To Einstein, a referee report was not an expected part of scientific publishing. It was an "incident" that deserved a sharp rebuke and a boycott of the offending journal. And indeed, Einstein would never again publish a research article in *Physical Review*.

But Einstein's reaction likely came as a surprise to Tate, because by 1936, refereeing *was* a common practice at journals in the United States. This chapter looks at the expansion of refereeing from the UK to the US. Like other US scientific institutions, US-based learned society journals hoped to emulate the success of their European counterparts, and the first US refereeing systems seem to have taken direct inspiration from refereeing systems in Great Britain. This chapter also discusses expert advice at funding bodies, focusing on the Rockefeller Foundation, one of the most important sources of research funding in the early twentieth-century US. Like learned society journals, the Rockefeller Foundation consulted scientific experts for opinions on what to accept and reject, but they did not do so systematically or view those opinions as a requirement for a Rockefeller grant. Furthermore, the Rockefeller Foundation placed a tremendous premium on the personal qualities of an applicant, and scientists were being consulted not only about the scientific merit of the proposal but also about the applicant's teaching ability, personality, and likelihood of overall career success.

US SCIENCE IMPORTS REFEREEING

To understand the development of refereeing in the United States, it's important to understand something that many modern scientists find counterintuitive: In 1900, US universities and scientific institutions were largely considered remote, underdeveloped backwaters, both by European scientists and by most Americans. US scientists scrambled to get copies of European journals and often preferred to publish their best work overseas. European doctoral degrees were considered more prestigious than their US counterparts. Leaders in the US scientific community dreamt of remaking US universities along the lines of German research universities, where professors were encouraged to advance scientific knowledge rather than focus most of their time on teaching. In 1883, the physicist Henry Rowland, president of the American Association for the Advancement of Science (AAAS), called US universities a "cloud of mosquitoes, instead of eagles" and criticized both the salaries offered to professors and the conditions of university laboratories.[3]

The US did have learned societies and was beginning to develop research universities. The National Academy of Sciences had been founded in 1863 amid the Civil War with a goal of providing scientific advice to the Union military; the American Association for the Advancement of Science (modeled on its British counterpart) was founded in 1848; Johns Hopkins University, the first research university in the US, opened its doors in 1876. All of those institutions, however, were conscious imitations of European institutions and were considered fledglings by the international scientific community. The same went for US scientific journals. *Science*, founded in 1880, was modeled on the British *Nature*, as its founding editor proudly declared in its first issue.[4] And the existence of US journals did not necessarily make them appealing places to publish, even for local researchers. The US radioactivity physicist Bertram Boltwood, for example, regarded European journals as far more significant for his work than American journals because the researchers he hoped would read and cite his papers almost all worked in Europe.[5] Through the middle of the century, most US scientists saw European science as an example they needed to catch up with and a model to emulate.

That emulation, as we shall see, extended to the adoption of refereeing at US learned society journals. In 1918, the chemist Arthur Becket Lamb assumed the editorship of the *Journal of the American Chemical Society* (*JACS*). Lamb would go on to edit *JACS* for just over thirty years, shepherding it through the disruptions of the Great Depression and World War II. He is best remembered for a particular editorial innovation: a referee system, the first of its kind at *JACS* and the earliest instance of systematic refereeing that I've found at a US journal. The response from *JACS* contributors was not particularly enthusiastic. As Lamb would recall in a 1952 history of *JACS*, contributors feared a delay between submission and publication, and referees sometimes "felt that it was underhanded" to send criticism to an editor behind an author's back.[6] Despite the objections, Lamb strongly believed his new system was an improvement over review by the editorial board. "The editor's aim is to publish the maximum number and the highest quality of contributions to the *Journal*," he argued. "To do so, he needs the objective opinion and advice of the best-informed and most sagacious experts in the field of each manuscript." However, said Lamb, both referees and editors needed to remember that "it is *his* [the author's] manuscript; he, and not the *Journal*, is responsible for the statements made."[7] The desire to ensure *JACS* articles were high quality, the emphasis on utilizing collective expertise, and the conviction that the author was ultimately responsible for the manuscript's contents would have been very familiar to George Gabriel Stokes a half century earlier.

Another useful example is that of *Physical Review*, founded in 1893. *Physical Review* was initially funded by Cornell University under the editorship of the Cornell professor Edward Nichols. The journal actually predated APS itself; APS was founded in 1899. When APS was first established, the new society and *Physical Review* were closely connected—the title page of *Physical Review* read "Conducted with the Cooperation of the American Physical Society"—but they were technically separate organizations.[8] That changed in 1913 when APS leaders persuaded the *Physical Review* team at Cornell to allow *Physical Review* to become the APS's official journal. APS wanted its own journal in order to bolster its reputation as a legitimate European-style scientific society, and bringing *Physical Review* into APS was both straightforward and avoided offending the Cornell contingent in APS by starting a competitor.[9] The journal continued to be run out of Cornell under the editorship of G. S. Fulcher until 1926, when Fulcher stepped down and Tate, a University of Minnesota physicist, stepped in.

Tate, who edited *Physical Review* from 1926 to 1950, turned out to be a transformative figure for the journal. He oversaw *Physical Review*'s ascension from a purely regional publication read by few outside the US to a journal with international reach. In Tate's National Academies of Science *Biographical Memoir*, former colleagues Alfred Nier and John van Vleck said that it was Tate's sophisticated understanding of quantum mechanics that brought *Physical Review* into the top tier of international physics journals. Credit also belongs to Madeline Mitchell, the sole member of the Minnesota *Physical Review* office staff, who was so effective in the role that in 1931 the newly created American Institute of Physics (an umbrella organization that supported publications for multiple learned societies in the physical sciences) made her its publications manager.[10]

One key to Tate's success, said van Vleck and Nier, was his willingness to get top work into print quickly. Nier and van Vleck recounted one story of Tate recopying an author's equations by hand because the printer could not read them and he did not want to delay publication by sending the paper back to the author for a clearer copy.[11] Tate's emphasis on speed of publication also sometimes meant little or no refereeing. "Tate showed rare judgment and common sense in not delaying by much refereeing noteworthy papers dealing with various applications of quantum mechanics," Nier and van Vleck wrote, adding that one major paper—David Dennison's "The Rotation of Molecules"[12]—was "refereed" through a quick in-person consultation with a Minnesota colleague in the *Physical Review* office.[13] That short anecdote gives us a good sense of what Tate thought refereeing was for at *Physical Review*. He was clearly willing to consult advice from physicists outside the *Physical Review* editorial staff, but he did not see formal reports as a crucial part of the process the way they were at Royal Society journals. Instead, his priority was getting innovative research into print quickly, with or without an external report.

When Tate did ask for a written report on a paper, it seems to have been because he thought the paper was of borderline quality or questionable in some way. To get a sense of how refereeing at *Physical Review* worked in practice, the historian Roberto Lalli has studied the contributions of one frequent referee: Howard P. Robertson, known for his research on quantum mechanics and cosmology. Tate first asked Robertson to referee for *Physical Review* in 1930, when Robertson was a twenty-seven-year-old assistant professor at Princeton, and Robertson quickly became one of Tate's go-to referees. If Robertson deemed a paper unworthy of publication, he did not mince words. To quote one of his most memorable comments, Robertson wrote that one paper could perhaps be improved "if it were written in invisible ink."[14] The authors of the papers Robertson reviewed did not see Robertson's most cutting comments. Instead, Tate would write a letter to the author explaining his editorial decision and sharing such referee comments as he felt appropriate.

Even though he must have had to take care in paraphrasing Robertson's harsh assessments, Tate did not object to Robertson's sharp takes on his colleagues' work. In fact, that very sharpness seems to be one reason Tate sought Robertson out. The default decision for a submission to *Physical Review*, at that time, was "accept"; as noted above, only papers that Tate felt were in some way questionable were sent out to a referee. Often, Tate's letters to Robertson reflected some personal skepticism about the article he was sending. "I can make nothing of this paper and I get the impression that it is mostly throwing one's weight around," he told Robertson about one paper in 1933.[15] After Robertson's first experience as a *Physical Review* referee, Tate thanked him for doing the "dirty work" of outlining just exactly what was wrong with a particular submission[16]—implying that Robertson had given Tate the answer he expected and had made his job easier by providing reasons to reject the paper. Nor was Robertson the only referee who sent blunt assessments to the editor. In one report from the early 1930s, the MIT physicist John C. Slater told Tate that "the paper which you sent me to review seems to be entirely crazy."[17]

Robertson did not savage every paper he was sent, however, and seems to have taken the stance that when in doubt, the decision should be "accept." Robertson also took an author's reputation seriously when deciding whether to recommend a paper for publication. Even when Robertson was critical of an article, he would suggest acceptance if the author was an established figure in the field. As he told Tate in one letter, an author "of good scientific standing . . . has more right to be heard than any single referee has to throttle!"[18] Reputation alone, however, was sometimes not enough to guarantee acceptance. It was Robertson who wrote the critical assessment of the Einstein-Rosen paper on gravitational waves. Robertson also used his status as a referee to place his thumb on the scale of ongoing scientific debates. For example, Robertson intensely disliked the English physicist Arthur S. Eddington's interpretation of special relativity, and he frequently recommended rejection for papers that seemed to endorse Eddington or that used a similar line of reasoning.

When evaluating an anti-Eddington paper by Einstein's former coauthor Boris Podolsky, Roberton deemed the submission "not too exciting" but argued that it was "justifiable to publish refutations of stuff published by supposedly reputable scientists as A.S.E."[19]

Robertson's approach to refereeing, and Tate's reliance on his opinions, suggests several important conclusions about the purpose of refereeing at *Physical Review*. Referees were most influential when the editor felt skepticism or concern about a paper; indeed, those were the *only* cases in which Tate asked for referee reports at all. The editor's "desk accept" was still a common decision. Referees came in handy when Tate did not feel he had sufficient expertise to make a judgment about a paper—or in cases where he felt rejection was the correct decision but needed help justifying that decision to the author. In the end, however, Tate felt that as editor he bore the primary responsibility for making decisions about what would be accepted and rejected. Tate consulted referees because he found them useful, not because he thought it was in any way a requirement.

PRIVATE SCIENCE FUNDING IN THE US

World War I proved a transformative moment for the US scientific community. During the war, leading figures in the US scientific community, including the astrophysicist George Ellery Hale and the physicist Robert Millikan, cultivated ties between US scientists, industry, the military, and the federal government, hoping to lend their scientific expertise to wartime scientific and technical problems. In 1916, the astronomer George Ellery Hale pledged the National Academy of Science's services to the government in the event of hostilities. That offer eventually led to the formation of the National Research Council (NRC), a group at the National Academy of Sciences that aimed to provide scientific services to the federal government.[20] Although wartime research did not have a major impact on the outcome of the war, the ties that US scientists forged gave them new political and economic power and made scientific advice to both the military and the government commonplace.[21]

Those new ties resulted in new prestige but did not lead to an outpouring of funds from the federal government. By the early twentieth century, however, US science had another source for scientific research funding. The US had an unusual provision in its tax code that made starting private charitable foundations financially desirable for the country's wealthiest heirs and families. Large foundations dedicated specifically to scientific and medical research began springing up in the early twentieth century. The Rockefeller Institute for Medical Research opened its doors in 1901, the Carnegie Institute for Science was established to support research in 1902, and the Russell Sage Foundation opened in 1907 with a mission to support the social sciences. A few decades later these would be joined by the John A. Hartford Foundation for medical research in 1929 and the interdisciplinary Ford Foundation in 1936.

Scientists in the early twentieth century tended to find these private foundations far more appealing sources of money than the federal government. Prior to World War II, most federal government funding for scientific research in the US was for applied research aimed at executing specific projects or creating particular technologies. Private foundations, on the other hand, often had broader aims, such as furthering knowledge or benefiting education. Carnegie's deed of trust to his foundation declared that the foundation's job was to "discover the exceptional man" and "enable him to make the work for which he seems specially designed his life work."[22] The upshot of that sort of lofty, vague language was that scientists could receive private funding for projects of their own design so long as they convinced the foundation of their scientific importance.

Furthermore, private foundation grants could dwarf those available from other sources. When it opened in 1902, the Carnegie Institute's endowment was ninety times that of the National Academies of Science.[23] The gap between government and private funding only grew during the Great Depression. Federal science funding plummeted during the early 1930s and was not restored by the large-scale spending of the New Deal due to both fiscal concerns and to a cultural backlash against technology during a period of high unemployment.[24] Private foundations also had to grapple with their endowments shrinking during the Depression, but their grant programs did not contract as dramatically as government ones did. Some federally affiliated science projects even found themselves applying to private foundations for money to support their work. In 1933, President Franklin Roosevelt created a new organization called the Science Advisory Board, charging it with the task of helping government departments with their scientific problems.[25] That help, however, did not include any financial aid. "The Board has not received any funds from the Federal Government," the NRC executive secretary Albert Barrows told the New Jersey congressman Charles Wolverton in 1935. "The Board has suggested to the President that large sums be devoted to scientific research, but these funds have not been appropriated . . . The Board has been supported wholly upon funds received from non-governmental agencies, mainly the Rockefeller Foundation."[26] The Rockefeller Foundation ended up giving the board a number of grants, ranging from a generous $50,000 initial appropriation to a small grant to publish their reports.[27]

Because of their flexibility and comparative wealth, private philanthropic organizations wielded enormous influence over US science in the early twentieth century, particularly basic science performed at universities. The various private foundations whose money shaped US science operated on different models. The Hartford Foundation, for instance, began as a shoestring organization run entirely by John Hartford, the founder of the A&P supermarket chain. Throughout the 1930s, Hartford personally chose grant recipients, with the foundation's board of directors serving as a rubber stamp for his decisions and little more.[28] Hartford's rationale for such a bare-bones operation was that money would be far better spent on medical research than on administrative costs. The Ford Foundation and the Carnegie Institute were both larger organizations. The Ford Foundation became known for its programs helping European scientific organizations rebuild after World War II and for its influence on the social sciences.[29] In 1910, the Carnegie Institute expanded into the Carnegie Corporation, a large general-purpose foundation that had its hands in whatever philanthropic efforts the trustees deemed desirable. The Carnegie Corporation would go on to be a generous provider of scientific fellowships, a major contributor to the budget of the NRC, and a provider of limited grants to universities for research.[30]

The most influential private science funder, however, was arguably the Rockefeller Foundation, which opened in 1913 as the successor to the Rockefeller Institute. John D. Rockefeller, the founder of the Standard Oil company, endowed the foundation with a staggering $182 million (roughly $6 billion in 2025 dollars) and a mission to improve global public health. In the 1910s and 1920s, most of the Rockefeller Foundation's work was focused on medical training and, occasionally, medical research, and most grants were made at the institutional level in large dollar amounts. Over the course of the 1920s, however, the Rockefeller Foundation became more and more interested in funding science that was not strictly medical in nature, and a significant reorganization during the Great Depression put emphasis on grants to individual scientists rather than departments or programs.[31] The Rockefeller Foundation's largesse is nicely illustrated by some figures from 1934. That year, the Council to the Administrative Committee of the NRC awarded a total of $16,497 (just over $400,000 in 2025 dollars) in grants to scientific researchers.[32] That was not a paltry sum, but in that same year the Rockefeller Foundation awarded $268,000 in fellowships in the US alone and appropriated over $1.3 million for its program in experimental biology.[33]

The figure who most shaped the Rockefeller Foundation's programs in interwar science was Warren Weaver, a mathematician from the University of Wisconsin. Weaver began working at the foundation in 1932 after being recruited by his former Wisconsin colleague Max Mason, the president of the foundation from 1929 to 1936. By the mid-1930s Weaver had been placed in charge of the foundation's programs in a wide range of biological and biophysical sciences.[34] It was under Weaver's guidance that the Division of Natural Sciences became arguably the single most important source of US science funding prior to World War II. The historian Robert Kohler has characterized Weaver as an "activist manager of science,"[35] someone who deliberately sought to use Rockefeller money to shape the direction scientific research took in the US. Specifically, Weaver was interested in the burgeoning research field of "vital processes," an interdisciplinary group of projects that focused on viruses, organic chemistry, biophysics, and what we would now call molecular biology.[36]

The records of the Rockefeller Foundation reflect grants to universities across the US, to schools large and small, to researchers both well established and at the beginning of their careers. They also reflect a process of awarding grants that consistently integrated advice from scientists but left enormous power to determine final outcomes in the hands of Rockefeller employees. The Rockefeller Foundation did not, in other words, employ a formal refereeing system; it used what Kohler has called a "patronage system," in which the positive opinion of a Rockefeller officer was the most important predictor of who would receive a grant.[37] Notably, the Rockefeller Foundation awarded grants all over the world and was particularly active in promoting research on public health in the developing world. For the purposes of understanding the role of expert opinions at a private funding body, however, this chapter will focus on grants for scientific research made to recipients in North America.

CHOOSING GRANT RECIPIENTS AT THE ROCKEFELLER FOUNDATION

The goal of a Rockefeller grant was "the advancement of knowledge"[38]—a mission statement that combined ambition with a large amount of flexibility. By 1940 the Division of Natural Sciences awarded a wide range of grants, from small grants-in-aid for hiring research assistants or purchasing equipment to fellowships for postdoctoral training to large-scale institutional grants meant to jump-start promising programs of scientific research. One common type of award that will be discussed in this chapter is a grant made to an institution to fund the research of a specific person, sometimes called a "fluid research" grant. Such grants were generally not transferable from one university to another, or from one researcher to another. If the researcher who had won the grant left the university where they had worked when it was awarded, the Rockefeller Foundation would generally consider the grant complete and would ask for a refund of the balance. Some researchers were able to move their Rockefeller money with them, but the foundation made it clear to applicants that grant transfers would be the exception rather than the rule.

Selection of Rockefeller grant recipients did not follow any one single process or method. Some grants began with a scientist contacting a Rockefeller employee about a project they hoped to have funded. Others resulted from a Rockefeller grant officer learning about a promising researcher and reaching out to them. Foundation employees generally preferred the latter method. Significantly, the Rockefeller Foundation was reluctant to have its funding acknowledged in papers written by scientists who received Rockefeller money. "I note that in these reprints you have acknowledged assistance from The Rockefeller Foundation," officer Frank

Blair Hanson wrote to one grant recipient in 1945. "The Foundation neither requests nor desires that such acknowledgement be given in papers representing work which has been in part supported the Foundation." Hanson acknowledged that this was "contrary to the policy of many such organizations," but he did not explain why the Rockefeller Foundation took a different path.[39] We can infer, however, that Rockefeller leaders did *not* want to spread the knowledge of their grant programs too widely among scientists. They strongly preferred to have contact between the foundation and a scientist initiated through personal introductions or their own initiative rather than an appeal from an unknown researcher.

Even when the foundation was the one to initiate contact, the scientists who hoped to receive the money were asked to write out a proposal indicating what the money would be used for and how it would advance scientific knowledge. The format of those proposals would look familiar to most modern scientists; the documents generally contained a narrative about the work to be done, a description of the methods used, a budget, and a list of publications by the potential grantees. Recipients of prior grants were also expected to account for how they had spent previous Rockefeller money.[40] However, a poor proposal did not necessarily sink an applicant's chances. Weaver deemed one 1939 proposal from geneticists at Columbia "rather vague and unconvincing" but added that the applicants "are both good men, and the subject is an important one which falls explicitly within our interests."[41] The Rockefeller Foundation would go on to fund genetics research at Columbia for nearly two decades, giving them over $350,000 (roughly $7 million in 2025 dollars).[42] As we shall see, the Rockefeller Foundation valued other factors—including the opinion of an applicant's fellow scientists—more than the ability to write a strong proposal.

Power to make decisions about funding technically rested in the hands of the Rockefeller Foundation trustees, including John D. Rockefeller himself, his son John D. Rockefeller Jr., the politician and Supreme Court Justice Charles Evans Hughes, the businessman Julius Rosenwald, the Rockefeller advisor Frederick Taylor Gates, and the former president of Harvard Charles William Eliot. The trustees had been deeply involved in decision-making about funding for scientific research when the foundation concentrated on making large institutional grants in the 1910s and 1920s, but when the foundation began emphasizing smaller grants for research projects undertaken by specific individuals during the Great Depression, it became impractical for the trustees to be true decision-makers. There were too many grants for the board to review in detail at its annual meeting, and the trustees had not been chosen for their scientific expertise. In practice, it was the funding body employees, called "officers," who determined which projects would receive money and which would not.[43] "Sole responsibility for recommendations on activities in the United States and Canada, and final responsibility for recommendations on activities in Europe and elsewhere, fall on the officers located in New York," Weaver explained in a 1938 document on administrative procedures.[44]

To make those recommendations, Rockefeller officers cultivated what the historian Michael Barany has called a "distinctive form of expertise, not in science or of science, but about science"—in other words, they built up knowledge not about scientific theories but about scientists and the places where they worked.[45] Officers would visit potential and current grantees in person, evaluating the conditions of their laboratories and their reputation among their colleagues. In a 1946 booklet intended for employees of the Division of Natural Sciences, Weaver reiterated that officers were entrusted with significant discretion over which grants would be funded. The booklet told employees to get to know applicants on a personal basis to determine if they were the kind of mind who was likely to have a major impact on US science. Rather than interview scientists, Weaver recommended that Rockefeller officers learn about their work and their background by "sitting on stools and smoking with them." Rockefeller officers should, however, do their best to downplay their individual influence; Weaver suggested they tell applicants that any funding would have to be the decision of a foundation "Committee." In reality, there was no such committee.[46]

A project's personnel and facilities were not the only factors Rockefeller officers considered. For example, officers often discussed whether a particular university or department would continue to support the project after the grant period or if the project would wither on the vine when the foundation money ran out. The potential for support from other foundations was also something they considered; the officers' diaries record many conversations with employees from other foundations about interesting scientists, and there seems to have been a deliberate effort to make sure promising applicants received funding while avoiding duplicate grants.[47] Rockefeller employees also worried about giving too many grants to the same universities and actively worked to spread their money to small colleges, state universities, and universities that did not yet have strong reputations in the sciences but were seen as a promising potential home for scientific research.

In deciding whether a grant had the potential to advance knowledge, the foundation explicitly considered not just the scientific worthiness of the project but the reputation and personality of the person undertaking it. As Weaver explained it in a 1961 oral history interview, "When a person's personal peculiarities become so marked, so objectionable, so unsocial that other individuals will not work with him, then I think this has to be taken into account."[48] Weaver was not the only foundation employee who felt this way. The officer's diaries contain a great deal of commentary on the personality, appearance, and sociability of different scientists. Excessive shyness or awkwardness was remarked upon negatively, as was overt boastfulness, disorganization, and perceived laziness or lack of ambition. The officers preferred candidates who were good at training the next generation of scientists and would note a lack of graduate students with disapproval. There was a strong preference for married men with stable family lives, and there was also (as we shall see) open skepticism of women in science.

The officers' diaries were candid about what they felt were shortcomings of potential Rockefeller grantees. For instance, Hanson's diary for 1938 recorded a promising interview with Harvard psychologist Morgan Upton, but Hanson changed his opinion after learning that Upton's position at Harvard was a temporary one, not permanent as he'd implied, and that Upton was recently divorced from his wife. "It seems clear," Hanson wrote, that Upton "was something less than candid in his interview."[49] That same year, Hanson complained that one scientist "mentions no research interest or project" and declared, "I would not consider [him] for fellowship; not enough maturity, energy, or clarity of mind."[50] Weaver's diary noted with disapproval that the plant cytologist Barbara McClintock was "sore at the world because of her conviction that she would have a much freer scientific opportunity if she were a man."[51] When the foundation was considering a grant to the biochemist Hans Clarke, the foundation's file on Clarke noted concerns that Clarke was "overworked," as well as "thoroughly kind hearted, and can refuse none a place in his laboratory; result—too many students and for this reason the quality of some of the juniors [sic] work is poor."[52]

Not all such notes were critical. Weaver's diary entry for November 22, 1932, deemed the Harvard chemist G. B. Kistiakowsky "young, attractive, modest and very keen"; two months later, Kistiakowsky's work on organic reactions received a grant for $45,000.[53] A note on the University of Maryland physiologist Magnus Gregersen, who later received a grant as well, opened with the following: "33; gray hair; gives appearance of maturity; attractive personality; A.B. Stanford, Ph.D. Harvard with Prof. Cannon."[54] The cytologist Gordon Carlson was deemed "worth investigating" because he had "an attractive personality and talks intelligently about his research."[55] Weaver called biochemist Gordon Alles "a fine, big, attractive figure of a man, with a remarkably straightforward and modest attitude."[56] A 1935 Hanson diary entry about a possible fellowship for the cytologist Harriet Creighton noted that "Miss Creighton probably ranks second to Miss Barbara McClintock as a plant cytologist, but is without Miss McC's personal shortcomings." Hanson deemed Creighton worthy of fellowship consideration.[57]

Qualities beyond a scientist's control were also considered fair game for evaluation. The comments about Jewish applicants, for instance, often included a judgment of whether the scientist "seemed" Jewish. "F. is not at all Jewish in appearance and has no objectionable personality traits," Weaver wrote of the chemist Marvin Fox in 1937. In a note later that year about the chemist Milton Harris, Weaver wrote that "the Jewish characteristics which [chemist] Treat Johnson delicately refers to in his subsequent letter do not come very much to the surface in this meeting."[58] This fixation seems to be tied to what Pnina Abir-Am has characterized as Rockefeller's "risk averse" approach to bestowing grant money during the Great Depression.[59] Rockefeller officers did not want to give grant money to scientists who were unlikely to go on to long-term success. They knew that antisemitism was a barrier for Jewish scientists; for example, in Hanson's diary for 1937, he wrote that the biochemist David

Glick "makes an excellent impression and were it not for the fact that he is Jewish, would undoubtedly secure a good university appointment."[60] However, Rockefeller officers did not believe it was their responsibility to challenge or lower that barrier. Instead, they viewed the possibility of discrimination or prejudice against an applicant as a reason not to award a grant.

Scientists of color were seldom considered at all. Officers rarely met with Black scientists, and Black scientists who did receive funding were almost always part of a larger grant to an historically Black college or university.[61] In his 1946 guide for his officers, Weaver explicitly said that since discrimination against non-white, non-Christian, and politically active scientists would blunt "the effectiveness of the candidate as a future leader in science," applicants from those groups should not receive equal consideration for funding. Along similar lines, Weaver's pamphlet instructed officers to be skeptical about funding women because "an attractive and intelligent female candidate for a fellowship has a time expectancy of scientific activity which is perhaps half as long" as her male counterpart's career. "We are essentially interested in training leaders for the future," Weaver argued, and so "a woman has to present a considerably stronger than average case to justify the extra risk."[62] In his 1961 oral history interview, Weaver reiterated his view that women were "not as good risks as men. This is simply a biological and cultural fact, and I don't think we gain anything by disregarding it."[63]

However, almost all undesirable qualities could be ignored if the officers deemed the science of sufficient quality. McClintock, who had drawn Weaver's disapproval for daring to state that sexism might be hindering her career, received Rockefeller support on the grounds that "it would be a scientific tragedy if her work did not go forward."[64] In a 1941 diary entry about his visit to Cornell's biochemistry program, Hanson wrote that James Sumner's colleagues reported that "S. and his group were simply terrible teachers; they have the dirtiest and messiest laboratories and offices on the campus . . . [they] kept no record of their research funds, never knew whether they were in the red or had a balance, and did not care. They just went ahead and got these remarkable research results."[65] Perhaps because of the concerning campus visit, in early 1942 the foundation wrote to several experts in biochemistry asking for the experts' opinions on Sumner's work. They received glowing letters in reply, deeming Sumner's research groundbreaking and his laboratory extraordinarily productive. Sumner would go on to receive over $100,000 in grants from the Rockefeller Foundation. In these two cases, the Rockefeller Foundation's risks paid off handsomely; both McClintock and Sumner eventually won Nobel Prizes, an ideal outcome for a foundation that hoped its grantees would rise to the top of the scientific hierarchy.

EXPERT ADVICE AT THE ROCKEFELLER FOUNDATION

Sumner's story highlights one of the major ways that Rockefeller officers gathered information about potential applicants: through communication with scientific experts outside the foundation. The detailed diaries kept by the officers record dozens of meetings per month with scientists. At some meetings, the officer would talk with the scientist about their own work; however, there were other conversations that centered entirely around what a prominent scientist thought of a younger man (and it was almost always a man) whom the foundation was considering for a grant, or which younger scientists the foundation ought to have on their radar. "C. said that [John] Kirkwood, now at Cornell, is an absolutely outstanding young man in physical chemistry," Weaver wrote in his diary entry for February 15, 1937, after a visit to the MIT physicist Karl Compton.[66] Similarly, after a conversation with the University of Michigan biochemist Howard Lewis in January 1934, Hanson wrote that "L thinks that Vincent du Vigneaud (32 years) is the real comer among the younger biochemists."[67]

When a scientist was previously unknown to the foundation, the officers would frequently solicit letters from scientists who worked in similar areas, occasionally those recommended by the potential fundee. In 1940, for example, the foundation was considering a grant to the organic chemist Francis O. Rice, who had just moved from Johns Hopkins University to the much smaller Catholic University of America in Washington, DC. Rice had impressed Weaver during their conversations, but the foundation still wanted further advice on whether he would be a worthy grant recipient. Hanson wrote to several chemists, including the Nobel Prize winner Harold Urey, to ask their opinion of Rice's work. Hanson's letters noted that Rice had mentioned the recipient's name as an expert in the field and asked the recipients to "write us confidentially your estimate of Dr. Rice as a scientist and add any comments you care to make regarding the value and significance of his research problems."[68] The letters were glowing, and Rice received his grant.

Sometimes, the foundation would cite their own lack of expertise in a particular field as a reason for soliciting letters about an applicant. In 1943, William Boyd at Boston University applied for a two-year grant in immunology. "Since The Rockefeller Foundation is not staffed in this particular field, we should appreciate very much your frank and confidential opinion regarding Dr. Boyd's standing as a research man and also your opinion on the value and significance of the research problems upon which he is working," Hanson wrote in his note to the external experts.[69] The absence of an immunologist on the Rockefeller staff may not have been the sole reason to ask for letters on Boyd's work; one of the experts, Rene Dubos, noted that famed chemist Linus Pauling was promoting theories of immunology that ran counter to Boyd's theories. Dubos strongly recommended funding Boyd because he felt Pauling's claims "are based upon very insufficient experimental evidence" and thought that it was "extremely important that the same problem be dealt with in some other laboratory by independent workers."[70]

The length and detail of the letters about Rockefeller applicants varied.[71] When the foundation asked for feedback about Sumner, for example, the letters in reply averaged just two paragraphs long and simply deemed Sumner one of the best in his field.[72] In other cases, writers went into more detail about why they were impressed with a particular researcher or why funding would be helpful. The letters about Rice's application, for example, went on at length about his work and emphasized that Rice had been able to build an impressive research program at Catholic University, subtly assuring the foundation that Rice had the facilities to carry out his plans despite working at a university with only a modest scientific reputation.[73] Similarly, when Hanson asked Lewis for his views on the work of the Oregon State College biochemist Roger Williams, Lewis emphasized the "lack of financial support and inadequate laboratory facilities at Oregon State" and said "I do not know of anyone who has so successfully carried out an important piece of work under discouraging conditions as Professor

Williams."[74] The University of Chicago biochemist F. C. Koch, a frequent advisor to the Rockefeller officers, spoke directly to Williams's need for funding, writing, "I am confident Doctor Williams has the ability to carry out this work but that he cannot do so without the scientific help and material support he has applied for."[75] Williams was ultimately awarded a three-year $20,000 grant to study the synthesis of pantothenic acid.

It might be tempting to see the external opinions Rockefeller officers consulted as functionally equivalent to modern refereeing. The closest analogue, however, might actually be the early nineteenth-century French *rapports* system that we saw in chapter 1. Rockefeller grantees tended to be younger scientists building their careers. Rockefeller officers were consulting department heads, Nobel Prize winners, and others at the top of the scientific hierarchy—people whom the Rockefeller officers considered to be the scientific superiors of the potential grantees, not their peers.

Crucially, many of the files about Rockefeller grant recipients contain no letters from scientists, only records of comments by foundation officers. It seems that not every application led the officers to solicit letters, and the officers were comfortable awarding grants based on personal conversations. Furthermore, once a scientist established a relationship with the foundation, additional outside evaluation of his work was usually not considered necessary. For example, when the endocrinologist and Rockefeller beneficiary Frederick L. Hisaw moved from the University of Wisconsin to a new position at Harvard University in 1935, the foundation did not collect letters about his work or his new facilities; the decision centered instead around the funding situation at Harvard and whether Rockefeller money was needed at the new department.[76] The foundation ultimately continued to support Hisaw's work through the rest of the 1930s. Similarly, applications for grant renewals were rarely accompanied by a search for outside opinions, although the foundation did expect a detailed account of all spending and any results achieved with the foundation's money.

CONCLUSION

By 1939, systems of consulting external experts were present at some US funding bodies and journals, but as the examples in this chapter show, soliciting formal reports was not yet the norm. Asking outside experts for referee reports was one tool among many that the funding body employees and editors in this chapter used to make their decisions. It is notable that many of the reasons for using a referee system had to do with social factors such as safeguarding the reputation of an institution or reassuring rejected authors that their papers had received a fair assessment. Ideas such as "truth" or "objectivity" generally did not enter into the conversation about early refereeing systems. Although learned society publications and private foundations wanted to avoid printing or funding anything that was obviously wrong, the standards for "wrong" often differed based on social factors—consider, for example, Robertson recommending that Tate accept a paper he thought was incorrect because of the author's "scientific standing." Along similar lines, in the next chapter we will see the future *Physical Review* editor Samuel Goudsmit recommend accepting a paper he thought was low quality because it would be "simpler" than arguing with the author.

Institutional accountability is a helpful lens through which to examine what referee reports were for in the early twentieth century.[77] Who did the people running journals and funding bodies report to; in other words, who expected an account of the publishing and funding decisions? What groups were editors and funding body employees trying to please or cultivate, and what would be the consequences if they failed?

Learned society journals like *JACS*, *Physical Review*, or *Philosophical Transactions* were accountable primarily to the members of their learned society. Those members were readers of the journal, contributors to its pages, and were also affiliated with the society whose reputation would be affected by what these journals published. The editors did not want to publish poor work that would negatively reflect on their affiliated society, but they were also concerned with providing rejected authors with reassurance that their submissions had been evaluated fairly and by someone with expertise in their field. That is a notable contrast to the for-profit journals discussed in chapter 1. Those journals were accountable primarily to their publishers, who wanted to turn a profit—and there was little about refereeing that suggested it might improve a publication's financial bottom line. Commercial periodicals were also accountable to the contributors and readers who provided them with content and subscription money. Editors of for-profit publications often decided that they could best meet the needs of those readers and contributors by *not* using referees. Lockyer, for example, wanted to make sure *Nature* maintained its reputation for speedy publication; Lodge thought his readers would be best served if *Philosophical Magazine* continued to accept papers outside the mainstream of modern physics.

Private foundations were accountable primarily to their trustees and their wealthy benefactors. The officers who awarded grants had to convince a group of prominent elites who had mostly not been trained in the sciences that the money they were giving to researchers was being spent well and was advancing the foundation's goals. So long as they could do that, they had enormous latitude to award grants based on whatever they felt appropriate, whether it was exciting research, the right pedigree, or the absence of "personal shortcomings." To a lesser extent private foundations were also accountable to scientists, who might have balked at being associated with an organization known for funding questionable projects or scientists with middling reputations. The Rockefeller Foundation officers, certainly, took pains to cultivate positive relationships with leading scientists and saw such connections as crucial to their success. But time and again the Rockefeller Foundation made it clear that the most important factor in receiving a Rockefeller grant was the judgment of foundation employees. Recommendations from fellow scientists were a good sign, but that was not enough; the officers had to believe that the applicant had the right personal qualities and the right institutional support.

One common goal runs through all of these reasons for using referees: the well-being of the institution using a referee system. In the early twentieth century, it seems clear that the purpose of a referee report was to help decision-makers at an institution; having something refereed was *not* seen as a guarantee of quality or truth. We should therefore not find it at all surprising that some institutions used referees while others did not, or that many institutions used referees on an occasional or case-by-case basis. If decision-makers thought referee reports might be helpful, they requested those reports. If such reports did not seem useful, editors and funding body employees didn't bother seeking them out. We will see this pragmatic, institution-focused attitude toward refereeing in the next chapter as well, as refereeing became a surprising solution to the labor problems created by post–World War II scientific proliferation in the US.

3

"NOT PLEASANT WORK"

THE EXPANSION OF REFEREEING AFTER WORLD WAR II

In 1955, the editorial board of *Science* had a problem. Since the death of *Science*'s longtime editor James McKeen Cattell in 1944, an editorial board composed of scientists from different disciplines had chosen which submissions to publish and had suggested revisions to improve accepted papers before they went to print. On the surface, the journal was thriving; *Science* was required reading for any American scientist and had an increasing international readership as well. But submissions to the weekly journal had skyrocketed in number since the end of World War II, and the members of the board complained "that the job of refereeing and suggesting revisions for hundreds of technical papers is neither the best use of their time nor pleasant, satisfying work."[1] The editorial board's workload, in short, was no longer sustainable.

The solution? Begin outsourcing the review of papers to external referees.

This chapter explores external refereeing practices in the 1950s and 1960s at both funding bodies and scientific journals in the United States and argues that the impetus for the expansion of refereeing in the US was the dramatic increase of federal funding for science. At postwar American funding bodies, specifically the funding agencies affiliated with the US government, refereeing was used as a way of collecting scientific advice. Both scientists and technocratic policymakers saw bringing in scientist-reviewers as a way to leverage the status of postwar science and secure the support of the scientific community for the new agencies.

At journals, refereeing practices began to expand in the early Cold War for a more practical purpose: to ease the workloads of editors, who were trying to cope with the deluge of submissions created by the expansion of US science. However, embracing refereeing led to its own set of editorial challenges. The length of time papers spent in the refereeing phase of publication became a frequent cause for concern among editors, publishers, and contributors, and scientific authors began to express frustration with referee reports that suggested rejection on "trivial" or unfair grounds. Perhaps unsurprisingly, the expansion of refereeing led to an expansion of criticisms about it.

"HOW NICE TO BE A PHYSICIST"

When World War II came to a close, American scientists enjoyed an unprecedented level of prestige in their home nation. The nuclear fission bombs created by the Manhattan Project were widely credited with winning the war, and advice from scientific experts on how to approach the new nuclear reality was widely sought-after both in and out of the halls of Congress. An article in *Harper's* famously declared that "no dinner party is a success without at least one physicist."[2] The US military and federal government, pleased with the return on the $1.89 billion they spent on the Manhattan Project, began planning new investments in scientific research.[3] In 1947 the physicist Arthur Roberts marked the cultural moment with a tongue-in-cheek song called "How Nice to Be a Physicist," depicting physicists lording their prestige over senators and thinking of grant money "only when [they] wonder how to spend it." But the song was not an uncritical celebration of physicists' new relationship with the federal government. "If you find a fact essential/Classify it confidential," Roberts warned in the final verse. "Never give/A second thought/The FBI's approval must be sought."[4]

Despite occasional concerns about the power and control the government might gain over science and scientists, the growth of spending on science continued with little pushback from the US scientific community. Between 1938 and 1953, federal spending on scientific research in the United States increased by a factor of 25, adjusted for inflation.[5] The major reason for that government generosity was the intensifying Cold War between the US and the Soviet Union. Military leaders and legislators believed that science and technology might make the difference between victory and defeat in the event of an armed conflict between the two powers. Policymakers in DC spoke of a need for a "standing army" of physicists ready to be mobilized in the event that the Cold War turned hot. A 1952 Bureau of Labor Statistics report declared that physics was "vital to the nation's survival" and that it should be a national priority to ensure "a continuing supply of new graduates" in the field.[6]

After the Soviet Union launched the Sputnik satellite in October 1957, beating the Americans into space, many legislators and members of the public grew alarmed by the Soviet Union's apparent lead in science and technology. That worry received high-profile validation in December 1957 when the Vanguard TV-3 rocket meant to carry the US's first satellite into space exploded on the launchpad. The result was even more government spending on science and a new focus on the number of scientists and engineers being educated in America. In 1958, Congress authorized $1 billion ($11.28 billion in 2025 money) just for graduate training in science and engineering.[7] The National Science Foundation's budget for fiscal year (FY) 1959 was $133 million, triple that of its budget for FY 1958 and miles away from the modest $225,000 it had begun with in 1951.

The size and scale of American science grew to meet the call for an army of scientists—and to match the funding opportunities. The number of undergraduate and graduate degrees awarded in all fields in the US grew in the 1950s and 1960s due to GI Bill funding and the large Baby Boomer generation entering college. But it was the natural and social sciences that experienced the most growth. The fastest-growing field was physics; the number of physics PhDs awarded in the US had tripled following World War II and tripled again in the wake of Sputnik, a growth rate that was twice that of doctoral degrees overall. Physics departments grew as well. At the University of Michigan, the physics department attempted to hire eleven new faculty members in 1955 alone, but amid hot competition from better-endowed universities and the private sector, they only succeeded in hiring seven.[8] The celebrated physics department at the University of California-Berkeley grew so large that it became impossible to find a location on campus large enough to accommodate a department picnic.[9] Psychology and engineering experienced the next largest growth; degrees in those fields grew at over one and a half times the average rate of degrees overall.[10]

PROPOSAL REVIEW AT THE NEW NSF

One major result of new federal spending on scientific research was the establishment of two federal agencies dedicated to funding scientific research: the National Institutes of Health (NIH) and the National Science Foundation (NSF). Both agencies would later play crucial roles in establishing refereeing as a central aspect of science, and the respective review processes they adopted give us a useful look at what refereeing at funding agencies might be used for in the mid-century. So let us turn our attention first to the NSF, a brand-new federal agency whose sole purpose was to fund grants for scientific research.

The foundation of the NSF is often linked to the work and vision of the Republican technocrat Vannevar Bush. The multitalented Bush had been the director of the Office of Scientific Research and Development during World War II, and as the war came to a close, he began outlining a vision for postwar science that would continue the federal government's funding for scientific research into peacetime. He did so in part out of dissatisfaction with a competing vision for research funding being promoted by the West Virginia Democratic senator Harley Kilgore, who wanted the government to award contracts for research projects that might lead to public benefits or advances in military power. Bush felt that Kilgore was overemphasizing the military applications of science and technology and not doing enough to support basic research meant to advance knowledge. He began to develop his own scheme.[11]

The result was Bush's July 1945 report *Science, the Endless Frontier*, in which Bush issued a strong call for continued federal spending on science. With European science devastated by war, Bush wrote, "New impetus must be given to research in our country. Such new impetus can come promptly only from the Government. Expenditures for research in the colleges, universities, and research institutes will otherwise not be able to meet the additional demands of increased public need for research."[12]

One key part of Bush's vision was a new "National Research Foundation" that would "support basic research in nonprofit organizations"—that is, universities—and "develop scientific talent in American youth by means of scholarships and fellowships."[13] Bush also hoped that the foundation would become a centralized agency for all federal science funding, including work for medical advances and national defense.

The National Science Foundation Act of 1950 was signed into law after five years of legislative wrangling. The new NSF would do many of the things Bush had hoped a national research foundation would do, including award scholarships and fellowships and give grants to "basic" scientific research—that is, research that had the potential to advance knowledge but did not have any obvious practical benefit. However, the act did not create the single centralized science funding agency Bush had dreamed of. Other agencies, including the branches of the military and the National Institutes of Health (which we will discuss shortly), continued to be a major part of the government science funding landscape. Indeed, the NSF's first director, Alan Waterman, actively worked to restrict the fledgling agency's responsibilities, reasoning that it could be disastrous for the NSF to take on too much too quickly. Waterman carefully disentangled the NSF from responsibility for developing science policy or directing R&D at other government agencies. Under his leadership, the NSF was established as an agency that existed to encourage scientific progress in the United States by funding education and research in the natural sciences. (The social sciences would be added to the NSF's scope only later, in 1957.[14]) Its first budget allocation of $225,000 (roughly $3 million in 2023 dollars) was a drop in the bucket of the national R&D budget, which was roughly $1 billion across agencies.[15]

Interestingly, *Science, the Endless Frontier* made no recommendations for how Bush's imagined "National Research Foundation" should choose between proposals. Waterman and the others who helped set up the NSF were more or less starting from scratch. As Waterman would later tell it, NSF leaders "started from the premise that, in its broadest sense, national policy for science is a matter primarily to be determined by the scientists themselves" and that the NSF should award support "after consultation with leading scientists in their respective fields."[16] From the outset, the NSF's proposal review process was designed to involve a range of experts from the fields that the NSF was going to support.[17] Proposals would go to the NSF employee responsible for the program in that field. If the proposal was not obviously ineligible, that employee would arrange to have the proposal evaluated by relevant experts in the field. Some proposals were sent out for "ad hoc" mail review, in which copies of the proposal were mailed to scientists who submitted their comments by return mail. Other proposals were evaluated by special panels of experts assembled in person in Washington. The choice of the reviewers, as well as the choice of panel review versus mail review, was left up to NSF employees.

We can get a sense of what early NSF mail review was like from the papers of William F. Meggers, a physicist at the National Bureau of Standards known for his work on spectroscopy. Meggers was a frequent referee of spectroscopy and optical physics proposals, and his personal papers contain several NSF referee reports from the 1950s and early 1960s. The surviving copies of Meggers's referee reports indicate that the NSF asked three questions about each proposal: first, whether the investigator had the appropriate scientific expertise to execute the proposed research; second, whether the proposal had sufficient scientific merit to deserve a grant; and third, whether the applicant's institution had the resources to support the work. Meggers's written comments generally averaged less than a page, and the surviving referee reports all recommend funding.[18]

Meggers, like George Gabriel Stokes a century before him, also used personal correspondence to share thoughts directly with the scientists whose work he reviewed. If Meggers thought a proposal was particularly good, he might write to the applicant to inform them he'd given a favorable report to the NSF. For instance, in 1959 Meggers recommended funding two proposals from Purdue University, one from K. L. Andrews and one from Robert W. Stanley. Meggers's report to the NSF called Andrews's plans "praiseworthy and probably unique" and described Stanley's qualifications as "outstanding" and the proposal's scientific merit as "of the highest order."[19] After submitting the reports, Meggers wrote to Purdue to share that he'd endorsed both proposals and hoped they would be funded.[20] "Thank you very much for endorsing our NSF proposals," Stanley wrote in reply. "Dr. Andrews and I both appreciate your interest on our behalf."[21] Similarly, in 1955 Meggers reached out to the applicant K. W. Meissner to say that he had "approved your NSF proposal very enthusiastically," although he felt that "your Budget is too small for such an ambitious program."[22] In his reply, Meissner admitted that he agreed the budget was too small, "but I was afraid that I would not get the grant if I asked for more."[23]

It is also clear from Meggers's correspondence that in the NSF's early years, he did not consider a close professional or personal relationship with the applicant to be a conflict of interest. In 1955, for instance, he was invited to write a report on a proposal by the MIT spectroscopist George Harrison.[24] Harrison was a longtime personal friend and a fellow member of the editorial team at the *Journal of the Optical Society of America*. He was one of the very few correspondents who called Meggers "Bill," and the surviving letters from Harrison to Meggers are littered with friendly jokes and frank discussions of scientific disagreements.[25] That relationship, however, was not mentioned in Meggers's report on Harrison's proposal. Meggers simply deemed Harrison's qualifications "outstanding" and said the NSF should fund the proposal because it had the potential to "produce something of exceptional value."[26]

Referee opinions from experts like Meggers were only one piece of the NSF review process. The NSF ultimately left final decisions about which grants to fund in the hands of NSF employees and leaders. To receive funds, the proposal would not only have to win the approval of its referees but would have to be positively evaluated by NSF staff members, program leaders, and ultimately the NSF's governing board. Referee opinions were explicitly *not* the determining factor for whether the NSF would award or decline funding. As one internal NSF document explained the system, "It is intended that the program official, taking into account all pertinent factors and information input, will make a final decision."[27] Furthermore, scientists who submitted proposals would not be given copies of their reviews, only a summary report prepared by an NSF employee stating the major reasons for the acceptance or rejection of a proposal. Referee reports were confidential documents for the internal use of the foundation—just as they had been at the Royal Society a century before.

A NEW KIND OF REVIEW: "STUDY SECTIONS" AT THE NATIONAL INSTITUTES OF HEALTH

The NIH was the successor to several older biomedical research organizations in the government. In the mid-nineteenth century the Marine Hospital

Service—later renamed the Public Health Service (PHS)—created a laboratory to study bacteriological theory. In 1930, a government act renamed that laboratory the National Institute of Health and gave the new institute the right to award fellowships and grants for public health research. The grants program was initially modest; large-scale funding for biomedical research was not a federal priority during the Great Depression.

However, during World War II the wartime Office of Scientific Research and Development (OSRD) began funding biomedical research projects on a far larger scale than the federal government had previously done. As the war drew to a close, the OSRD's Committee on Medical Research began to look for a way to transfer roughly $2.5 million in ongoing projects to a peacetime agency. The obvious candidate was the PHS, and more specifically, the PHS's existing grants programs at the National Institute of Health. A new Research Grants Office was established to award and administer extramural grants for exciting projects and fellowships for promising biomedical researchers. However, some federal health institutes, such as the National Cancer Institute, still maintained their own grant and fellowship programs, and legislators and PHS leaders soon began wondering if it might be more efficient to bring everything under one centralized roof.

The vision of a centralized federal agency for biomedical research funding was realized in 1948, when a provision of the National Heart Act created the National *Institutes* of Health, an umbrella organization consisting of member institutes like the National Cancer Institute, the National Institute of Mental Health, and the new National Heart Institute. The new NIH oversaw all health-related federal institutes and their grant programs, and as part of the Heart Act, those grants programs underwent dramatic expansion. The NIH's total funding went from $3.4 million in 1946 to $28.5 million in 1949—an eightfold increase in just three years.[28] Bush's *Science, the Endless Frontier* inspired some of the vision for the new NIH, as did the example of the Rockefeller Foundation's programs for research and public health grants.[29] The most influential figures in the NIH's early years were the NIH's first director Rolla Eugene Dyer, an accomplished clinical researcher and former chief of the Division of Infectious Diseases, and the Surgeon General Thomas Parran. Dyer and Parran shared a goal of not just establishing the NIH's place in the wider federal system but enrolling the expertise and support of academic scientists.[30]

The NIH's Division of Research Grants created a system for reviewing grant applications inspired by the expert committees smaller grant programs at the NRC and National Cancer Institute had used before the war.[31] Each application went first to a small "study section" composed of scientific experts in a particular field. Members of study sections were generally distinguished researchers with established national reputations, and they would come to Washington, DC, for a week or more at a time to review applications and discuss which projects they thought should be funded. The study sections' recommendations were forwarded to an NIH council of scientists and laymen, which added their own recommendations. Final decision-making power rested in the hands of the institute directors, the heads of NIH member institutions such as the National Cancer Institute and the National Eye Institute. Although the directors took the earlier evaluations into account, they were not obligated to follow the recommendations of the study sections or the council.

The NIH's review system was, thus, different from any system we have discussed so far. To succeed, an NIH proposal would have to go through two sets of reviewers plus the institute director. The two stages of review—one by the study section, one by the council—had fundamentally different purposes. The study sections were supposed to evaluate proposals based on scientific merit. The council, on the other hand, evaluated them for a match with their institute's mission. The members of the study sections and councils were similar to referees at early twentieth-century learned society periodicals in that they were a relatively small group of evaluators and the scientists were mostly senior people in the field. However, in theory a learned society periodical could reach out to any number of experts; that was the model the NSF had adopted as well. In contrast, aside from a handful of paid ad hoc reviews solicited for unusual proposals, the NIH deliberately limited themselves to appointed study section and council members—and the council members did not necessarily have scientific training. Although the reviewers had formal NIH appointments, they were different from the Rockefeller Foundation officers or government employees who had evaluated grants in the early twentieth century because they were not NIH *employees*; each of them had full-time employment elsewhere.

The kinds of reports NIH reviewers produced were also quite different from anything generated at other organizations. The NIH convened study sections in person and expected discussion between their expert reviewers, and they did not assign specific reviewers to specific proposals.[32] Instead, the study section as a whole was supposed to make recommendations jointly. The councils took a similar approach, issuing recommendations to the director as a united body rather than attaching individual opinions to proposals. NIH applicants would receive little information about why their grants had been accepted or rejected. Deliberations about the grants were considered confidential and internal to the NIH. However, it seems that the minimal feedback was not initially a problem for most NIH applicants, perhaps because early success rates were quite high. In FY 1947 and 1948, for example, just under 75 percent of proposals were approved.[33]

The NIH system seems to have been built on the premise that the best scientific advice would come from a select board of highly qualified experts rather than ad hoc requests for reports from many different scientists around the country. The NIH described study section membership as a public service, a role in which scientists would offer their expertise in the service of their nation with a goal of strengthening US biomedical research as a whole. The goal of the NIH's review process was to choose research projects that had scientific merit, were likely to succeed, and furthered the institute's mission—goals the NIH's founders felt could best be achieved with input from a limited pool of senior US scientists. The lofty cultural status of science in the late 1940s made soliciting this kind of expert advice largely uncontroversial; there seems to have been little, if any, pushback to either the NIH's study sections or the NSF's use of referees.

THE INCREDIBLE EXPANDING SCIENTIFIC JOURNAL

The US government's science spending also transformed US science publishing—and not necessarily in ways that publishers or editors relished. More scientists meant more laboratories being built, more experiments being performed, and more scientific papers being written—and American journals struggled to keep up. Specialist journals across research fields experienced explosive growth in size. In 1950 the *Annual Review of Psychology*'s first volume contained 310 pages; by 1970, the size of a volume had more than doubled to 628 pages. The weekly *Science* printed 1,356 pages in 1945; in 1970, that number stood at 4,762. The *Journal of the American Chemical Society* went from 2,283 pages per year in 1945 to 7,632 pages in 1970; the *American Journal of Medicine* grew from 714 pages in 1946 (its first year of publication) to 1,656 in 1970.

If growing journal size had been a competition, however, the winner would have been *Physical Review*. In 1945, *Physical Review* published two volumes totaling 654 pages. By 1955, one year of *Physical Review* consisted of four volumes occupying 7,293 total pages. The journal's editor Samuel Goudsmit, who took over from Tate in 1950, began to worry that no one would actually read such a massive journal. In 1958 Goudsmit introduced a new journal, *Physical Review Letters*, dedicated to short articles from across all fields of physics. Goudsmit hoped the new journal would give physicists a place to publish their most exciting work as quickly as possible and lessen the pressure on *Physical Review*.[34] The new journal was a hit with physicists but did little to stem the growth of *Physical Review*. In 1969, *Physical Review* printed a whopping 24,533 pages, and that was the last year APS tried to publish their flagship as one unified journal.[35] In 1970, APS split *Physical Review* into four new journals: *Physical Review A* for atomic, molecular, and optical physics; *Physical Review B* for condensed matter physics and materials science; *Physical Review C* for nuclear physics; and *Physical Review D* for particle physics, gravitation, and cosmology.

Journal editors and publishers were not universally delighted with the new demands on their pages. One large publisher of journals in the physical sciences was the American Institute of Physics (AIP), an umbrella society that included APS, the Optical Society of America, and the American Association of Physics Teachers among its "member societies." In 1958, AIP[36] issued a statement about "Research on Publishing Problems in Physics" to be printed in their flagship magazine, *Physics Today*. AIP had anticipated problems in publishing after the end of World War II, the unsigned piece explained; "publishing costs were rising" and "printing and other production bottlenecks were commencing to be troublesome." But it turned out that this was no temporary blip. "Roughly 8000 additional yearly journal pages later, the Institute's publishing problem in physics has unmistakably become a serious matter," the release went on. Given the "relentless multiplication" of specialist scientific journals, the release said, there were problems with everything from timely publication to the availability of good abstracting services to shortages of typists who were capable of properly typesetting mathematical formulae. AIP announced that it had obtained an NSF grant specifically to research how to move forward under the new demands being placed on scientific publishers; the release expressed concern that "the Institute's future publishing program may be in jeopardy unless a solution can be found."[37]

NEW REFEREE SYSTEMS AT JOURNALS

As the *Physics Today* announcement suggests, the rapid growth in the number of scientific papers placed enormous strain on the publishers and editorial staff of scientific journals, and many journals had to reimagine their publishing process in order to keep up and survive. One example of a journal that sought to ease editorial burdens in this period is the *American Journal of Medicine* (*AJM*). When *AJM* was founded in 1946, its editor Alexander Gutman worked to ensure fast publication by handling the vast majority of acceptances or rejections himself. *AJM* became known for both its distinctive green covers (hence its nickname, the *Green Journal*) and its rapid submission-to-decision time of just seven to fourteen days. However, as *AJM* became more popular, Gutman was unable to keep up with the number of submissions. Slowly, the average time between submission and decisions grew to months rather than weeks. By the mid-1960s submissions to *AJM* were being sent out for external refereeing to ensure that the journal could maintain its reputation for fast publication.[38]

Another example of a journal that turned to referees is *Science*, the aforementioned weekly scientific journal that published articles from across scientific fields. In 1950, *Science* occupied an interesting place in the American science publishing landscape. It was technically a learned society periodical, affiliated with the century-old American Association for the Advancement of Science (AAAS). However, the magazine's weekly publication schedule and inclusion of both technical articles and the latest scientific news were qualities more often seen in for-profit publishing, and *Science*'s editorial board viewed the European commercial journals *Nature* and *Naturwissenschaften* as their most direct analogues.[39] Furthermore, under the leadership of its longstanding editor James McKeen Cattell, *Science* had operated far more like a commercial periodical than a learned society journal, heavily influenced by the ideas and attitudes of its strong-willed editor.[40] Indeed, until his death in 1944, Cattell was both the editor and the legal owner of *Science*, with the AAAS acting as a sponsor but exercising little control.

Cattell's death inaugurated a chaotic period for *Science*. The AAAS took over its legal ownership, but editorial stability proved elusive. The first AAAS editor, Willard Valentine, died in 1947 just a year after taking the job. As the AAAS searched for his successor, the AAAS Editorial Board member George Baitsell stepped into the role for a year. When Baitsell stepped down, the AAAS decided to place editorial responsibilities in the hands of the AAAS Editorial Board chair. Between 1949 and 1956 seven more people became de facto *Science* editors-in-chief, including the geologist Howard Meyerhoff, *Science*'s former administrative secretary; Bentley Glass, a prominent geneticist also known for his regular column at the *Baltimore Evening Sun*; and Dael Wolfle, a psychologist and longtime AAAS officer.

Not all of *Science*'s leaders during this period bore the title of editor-in-chief. Most were designated as "acting editor," "Chairman of the Editorial Board," or "Acting Chairman."[41] A string of female assistant editors and executive editors, including Mildred Atwood, Ruth Christman, Gladys Keener, and Beth Wilson, shouldered much of the burden of ensuring that the journal continued to publish regularly.[42] The editorial office, however, was woefully understaffed. A 1951 report to the board explained that *Science*'s staffing situation was "too tight for such forseeable contingencies as vacations, or such emergencies as prolonged illness."[43] In 1953, the new Acting Chairman Glass reported that there was "chronic low morale due in part to overwork" and that his predecessor Meyerhoff and former executive editor Keener had not been able to take vacations for over two years.[44]

Meanwhile, the average time between submission and publication for an article in *Science* was growing longer and longer. Editorial board members became increasingly frustrated with the manuscript backlog at *Science*; the May 1950 board meeting minutes recorded that the board voted to express "its profound dissatisfaction with the backlog situation."[45] That expression of dissatisfaction had little effect. Three years later, *Science* had a backlog of seven months' worth of accepted papers despite a roughly 50 percent rejection rate—unusually high for the time.[46] The editorial board voted to impose strict limits on the length of articles in *Science*: 1,200 words for technical papers, or one page of *Science*, in the hope of printing twelve to fifteen technical papers per week instead of just six to eight.[47] The plan apparently worked too well. A September 1953 letter from Acting Chairman Bentley Glass to the rest of the editorial board announced that the backlog "has been reduced to little more than one month's supply of technical papers, and this means that we do not have the necessary working margin because authors are not always prompt in returning their proof." Glass indicated that *Science* would need to begin accepting more papers to remedy the situation.[48]

Despite the dramatic shift in the backlog situation, it was becoming clear that a reputation for delay between submission and publication was discouraging scientists from submitting their work to *Science*, and that the board could not solve that problem unless they found a way to shorten the amount of time between the submission of an article and a decision on acceptance or rejection. During the messy post-Cattell period, *Science*'s editorial board handled almost all of the refereeing of *Science* papers. Each member of the roughly eight-man board (and it was all men) would review between two and four articles per month. Most articles were either accepted or rejected outright, but board members would also offer suggestions for corrections and improvement on articles that were deemed potentially suitable for publication. External referees were employed primarily for papers that fell outside any board member's expertise, or when authors were likely to object to rejection based on a single review.[49] Generally speaking, if a paper was ultimately rejected, the *Science* editorial board preferred to have at least two referees.[50]

The editorial board did not see writing their reviews more quickly as a viable solution. They felt that they were already working as fast as they could, and as we saw at the beginning of the chapter, they found the work tedious at best. In 1953, the editorial board proposed reaching out to members of the AAAS Council and Section Committees to act as "an extended panel of advisors for articles." That proposal does not seem to have gained much traction, however, possibly because council members were not terribly interested in taking on the job.[51] Meanwhile, *Science* urgently needed more material for its news and commentary sections. Most scientific journals did not have such sections, but *Science*'s board firmly believed that quality science reporting and commentary were crucial to securing subscribers.[52]

With AAAS officers unwilling to help, the board had to look for other solutions to the publication delay problem. It seemed clear that the board would need to look beyond the AAAS for refereeing labor. In 1953, the biologist W. D. McElroy wrote the board a letter suggesting that "the panel of referees should be much larger in order to cover a great variety of fields and to get prompt action."[53] In 1955, Glass (who had stepped down from his role as acting chair and was now simply a board member) recommended that the members of the editorial board should instead act as contributing editors, and that their primary responsibility should be to generate content for the magazine's news and commentary sections. That would involve reconstituting the editorial board so that it was composed of "people who could write and digest expertly and cover more fields of science," but it would be a small price to pay for solving *Science*'s persistent issues with backlog and board overwork.[54]

When the Stanford embryologist and university administrator Graham DuShane began his tenure as *Science*'s editor in 1956, he adopted Glass's model for distributing editorial responsibility. Editorial board members became contributing editors who oversaw the news and correspondence sections, and responsibility for refereeing was officially reassigned from the board to external referees, although board members continued to be consulted when a technical article fell into their field of expertise.[55] DuShane's efforts bore fruit. By the time he stepped down in 1961, the average time from submission to decision was down to just over two months.[56]

In 1962 the physicist Philip Abelson accepted the job as *Science*'s editor-in-chief. Abelson and the editorial boards who served with him did a great deal to develop consistent standards for refereeing in *Science*. By 1966, *Science* was consulting four thousand referees per year for both solicited and unsolicited articles.[57] Assistant editors would call potential referees by phone to give them information about a paper; if the referee agreed to review the paper, *Science*'s staff would send them a copy. Scientists who declined invitations were asked to recommend possible referees.[58] Each referee was sent a standardized instruction sheet indicating the criteria on which papers were to be judged.[59] Referees were asked for their opinion on the importance of the findings, the soundness of the methods, the organization and writing of the paper, the use of figures and illustrations, and whether other researchers had been treated fairly in the discussion and references. To be published in *Science*, referees were told that a paper "should be of unusual interest to the specialist or of broad interest as an interdisciplinary problem."[60] Significantly, instructions to referees indicated that most of the articles printed in *Science* had been solicited by the editorial team: "Most of our articles are solicited, usually on the recommendation of distinguished scientists. An unsolicited article should be given a fair review, and if referees agree that the material is excellent, it should be published."[61]

By the mid-1960s *Science*'s acceptance rate had been dramatically reduced from 50 percent to 30 percent, which did a great deal to ease the backlog. The acceptance rate decreased even though, as Abelson put it in a 1966 report to the AAAS Board of Directors, "Few of the papers submitted are considered unpublishable by our referees;" rejection was more often driven by "financial considerations and the need to keep Science at a reasonable size."[62] Despite the higher rejection rate, the February 16, 1968, editorial board minutes reported that "the staff has been quite successful in minimizing tensions with authors. Whenever an author tells us that his manuscript has been unfairly judged, the manuscript is cheerfully sent to additional reviewers."[63] Abelson also began allowing scientists to list potential referees whom they felt would be qualified to review their work. However, as Abelson would dryly tell his editorial board in the early 1970s, "Authors complain equally about prejudice or incompetence of referees, whether the referees are their choice or ours or a mix."[64]

Science's story nicely illustrates how the mid-century expansion of scientific journals could lead periodicals to adopt external refereeing when they had previously used other systems to decide which articles to print. Aside from one complaint about a paper that an editorial board member thought had not deserved publication,[65] there is little indication in the *Science* archives that the *Science* editorial board thought refereeing was necessary to improve the journal's contents from a scientific point of view. Instead, the emphasis was on editorial board workload and the speed with which decisions could be made about acceptance or rejection. Similarly, Abelson's later work to standardize refereeing practices was framed as a way to minimize author complaints, not as a process that was supposed to make the journal itself better. It's notable here that contributor unhappiness—with the long time between submissions and decisions, and later, with rejection—was a major feature of the discussion about refereeing at *Science*. The magazine's leadership was very conscious of keeping its contributor base happy and wanted *Science* to remain a desirable place to publish.

REFEREEING AT THE AMERICAN INSTITUTE OF PHYSICS

Refereeing had a longer history at the journals published by the American Institute of Physics (AIP), including *Physical Review* as we saw in chapter 2. By the postwar period, most papers submitted to an AIP journal received at least one referee report before publication, although those referee reports were often short, even perfunctory. In September 1949, for example, Samuel Goudsmit reviewed a paper for *Physical Review* and sent just 260 words of commentary to then-editor John Torrence Tate. "I don't know what to do about this paper," the report opened candidly. "Personally, I don't believe a word of it, but it's hard to put my finger on the weak spots. . . . Perhaps the simpler thing to do is just to publish it, as it seems hopeless to get into an argument."[66] The paper in question seems to have been published five months later.[67]

But editors of journals like *Physical Review* would not be able to speed a paper to publication in order to get rid of a stubborn author for much longer. During the publishing boom of the 1950s and 1960s, AIP answered the expansion in submissions to their journals by adding pages to established journals like *Physical Review* and by founding new journals like *Physics of Fluids* (launched in 1958) and *Journal of Mathematical Physics* (founded in

1960). Both the expansion and the new journals were aided by a 1961 shift in federal funding policies that made page charges for publishing in a scientific journal an allowable research expense for federal grants.[68] But AIP and the editors running its journals still needed to find ways to make editorial workloads manageable, and like the editorial board of *Science*, they saw refereeing as a potential solution. The new AIP journals of the 1950s and 1960s had refereeing systems from the outset, and the "desk accept" on an editor's authority became less and less common as editors outsourced the reading of papers to outside experts. In 1959, AIP's journal editors told the AIP Publication Board that "no manuscript is exempt from refereeing."[69]

However, one complaint about refereeing at AIP journals recurred again and again in the publication board's meetings: its impact on speed of publication. In 1960 François Frenkiel, editor of the new journal *The Physics of Fluids*, wrote to James Krumhansl, the Chairman of AIP's publication board, on the subject of "how the quality of scientific journals can be measured and what should be done to continue to improve the AIP publications." Frenkiel admitted that he was unsure "whether there is a common measure by which the quality of scientific journals can be determined," but he did suggest a list of ten factors that he felt readers and contributors took into account when thinking about whether or not a scientific journal was high quality. Alongside criteria such as "content and scientific quality of papers" and "nature and number of readers," Frenkiel included "publication delays of papers and reprints."[70] This was an area of concern for the AIP board as well. Every year the board asked editors to collect data on how long it took an accepted paper to go from initial submission to publication, and it seemed that the board was growing increasingly frustrated with the answers—and with journal issues that missed their targeted printing dates, sometimes by a month or more.[71]

Refereeing was one of the obvious steps at which publication was often slowed down. For the publication year 1962, the time that published articles spent being refereed ranged from just over ten weeks at *Review of Scientific Instruments* to over twenty weeks at the *Journal of the Optical Society of America*.[72] When the matter came up for discussion at meetings of AIP's publication board, however, there seemed to be little hope for change. In 1959, editors assured the board that they provided referees with self-addressed stamped envelopes to return the manuscripts and their comments, but they balked at the suggestion of introducing a standardized check list to speed the process.[73] Further discussions seemed to result in little movement on the issue. Faster refereeing "would be welcomed by the scientific community," the board agreed in 1962, but "the editors present saw no way of improving notably over present performance."[74] "We cannot complain about referees," *Physical Review Letters* managing editor George L. Trigg said in his report to the board for 1962. "The service they render is invaluable, and they perform it well and readily. Most authors, likewise, recognize the value of reviewers' comments, and many have taken the pains to tell us so."[75]

So it seemed that both the AIP board and AIP journal editors regarded refereeing as an important piece of the editorial puzzle, one that was necessary and could not be rushed. But what did they think refereeing was *for*? Why did they view it as worth doing, despite frustrations over the length of time papers spent with referees? In 1963, editors of each AIP journal submitted copies of the instructions they sent to referees as part of a board examination of refereeing practices at AIP journals. The resulting collection of instructions to referees provides an extremely useful window onto what referees were being asked to consider—and, therefore, onto what editors were hoping their referees would do for their journals.

Three research journals gave minimal instructions to their referees.[76] The *Journal of Mathematical Physics* sent only a short cover letter asking the recipient to "kindly review this paper and advise of its suitability for publication."[77] The *Journal of the Optical Society of America* asked referees to give the editor their opinion on its "originality and merit."[78] *The Journal of the Acoustical Society of America* also did not send referees a form: In a handwritten note to Hugh Wolfe, the director of AIP's publications division, the editor Bruce Lindsay, said that his journal "has no official form" for referees; "each associate editor is free to ask for comments in the form most convenient for his purposes."[79]

The rest of the journals used a standardized list of questions, ranging in length from just three queries (*Applied Optics*) to eleven (*Physical Review*). Common questions included:

- whether the paper contained new findings interesting enough to merit publication
- whether the paper was suitable for this specific journal

- whether the paper's length was appropriate
- whether the paper cited references appropriately
- whether the abstract was well done
- whether there were errors, especially in the paper's mathematics

Most editors asked referees to return papers within two weeks, with the exception of the fast-publishing *Physical Review Letters*, which asked for a one-week turnaround.[80]

Several of the referee questionnaires included specific questions about the writing and organization of the paper. *The Journal of Chemical Physics* asked referees to weigh in on the quality of the paper's English and the length of the paper's sections, as did *The Astronomical Journal*.[81] Similarly, the 1963 form for *Physical Review* included questions about whether the paper was "written intelligibly," whether the figures were easily understandable and properly captioned, and whether the English could be improved, with a request for referees to "make suggested changes in pencil in the margin of the manuscript."[82] Just a year earlier, however, the managing editor Simon Pasternack had flagged those areas as things that *Physical Review* referees *didn't* usually do, or at least not very well. Referees, he wrote in his report for 1962, "don't usually consider" issues such as "minimally decent English, clearly arranged tables and figures, consistent notation, properly specified units, elimination of defects such as incomplete or improper references, obvious duplication, excessive claims of precision, [or] unjustifiable priority claims." Those tasks, said Pasternack, were really the province of the editorial staff, and specifically the trained physicists on the staff, since "copyreaders who have little or no background in physics" could not be expected to solve such problems.[83]

Referees, then, were clearly not expected to be much help with the editing of articles, even if they were specifically asked to provide feedback on style, language, and formatting. Instead, it seems that editors of journals were primarily relying on their referees for help with determining the paper's significance. A question about the novelty or originality of the scientific work in the paper appeared on every single referee form. In an era of exploding journal lengths, editors were understandably eager to identify and publish the most significant papers rather than print results similar to ones already presented elsewhere. Familiarity with the existing literature in a specific area of expertise, not skill as a copy editor, was the major desired quality of a referee for an AIP journal. In these documents, we also start to see AIP editors begin to articulate the idea that referees improved the quality of a journal—an idea that was notably absent from earlier conversations.

It is also worth noting that by and large, aspiring contributors could expect that their papers would eventually be accepted. The usual rejection rate for an AIP journal hovered around 15 to 20 percent, meaning acceptance was more likely than not even at a high-profile journal like *Physical Review*.[84] Referees helped by separating out the small number of papers that were clearly unsuitable, not by harshly judging every paper and deeming only a select few worthy of publication.

"THE PRESENT SYSTEM IS ROTTEN TO THE CORE"

Other AIP Publishing records reinforce the impression that by the early 1960s, refereeing was an expected part of scientific publishing in the United States. None of the instructions to referees discussed above, for example, explained what a referee report was or seemed to anticipate that scientists might be unfamiliar with what they were being asked to do; in fact, refereeing was often described as an obvious scientific duty. "We would like to re-iterate the opinion that refereeing an article in one's own particular field should be regarded as a professional obligation: none of us should be too busy to do this at least occasionally," the *Applied Optics* editor John Howard wrote to prospective referees in 1963. "The excellence of a journal is determined by the overall quality of the individual manuscripts, and the Editor is guided in his assessment of the manuscripts almost exclusively by the reports of the readers."[85]

The expansion of refereeing was most dramatic in the US, but it's worth pausing to remember that the US was not the only country where journals might employ referees. Refereeing was also still an expected part of scientific publishing at learned society journals in the UK. At the Royal Society of London, for example, referees were still a standard part of publication. In 1968 the society would abolish the rule that only Fellows could be referees for society journals, significantly expanding their pool of referees in a bid to speed up the steps from submission to publication.[86] When the physicist John Maddox became editor of the for-profit British journal *Nature* in 1966, he began gathering reviewers in person to evaluate submitted papers, although the dynamic Maddox also felt it was an editor's privilege and duty to accept some papers on his own authority.[87]

Complaints from the 1960s, however, made it clear that scientists viewed Americans as unusually reliant on refereeing—and that not everyone agreed this was a positive quality. In 1962, the chemist Howard K. Zimmerman told the physicist Jacob J. Bikerman that he found American journals intolerable because of the refereeing system. "There is a considerable number of scientists in this country who are more or less intensely disillusioned about American editorial policies," he wrote in a letter to Bikerman. "I think our editors are either incompetent or afraid to make their own decisions about what papers to print and what to reject." The reliance on referees, he argued, led to slow publication and "theft of scientific ideas," and enabled editors to choose reviewers they knew would savage a paper. "The present system," Zimmerman declared, "is rotten to the core." Zimmerman had therefore made the decision "to publish all my work that I consider to be of any consequence in foreign journals."[88]

Bikerman, a Soviet émigré who had come to the United States during World War II, was an experienced referee.[89] However, by the 1960s he seems to have agreed with Zimmerman that the American system was flawed. In his reply to Zimmerman's letter, he wrote that Zimmerman's sentiments were "very similar to mine" and added that he did not buy the claim that referees made a journal better. "Papers in Kollod-Zeitschrift [a German journal of colloid chemistry] are about as poor as the colloid chemical papers in J. Phys. Chem.; Thus, having referees does not raise the scientific level of a journal." The American *Journal of Physical Chemistry* seems to have been a particular target of Bikerman's ire. He told Zimmerman that he planned to collect names of irate scientists and to notify the editor William A. Noyes Jr. of "every instance in which a paper rejected by him was praised and confirmed by other scientists."[90]

Bikerman and Zimmerman were not alone in their frustration with refereeing as a system. "The present refereeing system penalizes articles advocating unconventional ideas," theoretical physicist Michael J. Moravcsik wrote in a 1965 essay for *Physics Today*. Moravcsik suggested that authors should be allowed to have their paper published alongside a negative referee report rather than accept rejection.[91] In addition to "stimulating further thinking," Moravcsik argued that this option would improve the quality of referee reports, since referees would know that their report could be made public. "I am fairly certain that a quick opinion survey among physicists would immediately disclose that there is a need for such a rise in standards, particularly in connection with the *Physical Review Letters*," he wrote, indicating particular frustration with the sought-after venue for quick publication. It is worth noting that *Physical Review Letters* had a 50 percent rejection rate, well above the average 20 percent at other AIP journals, which is likely why Moravcsik singled it out for criticism.

CONCLUSION

In the two decades following the end of World War II, refereeing expanded substantially in the United States. At journals, the expansion was largely a response to increased workloads; at federal funding bodies, advice from scientists was seen as a way to enroll scientific expertise and cultural prestige in service of new agencies. All of this meant that refereeing was becoming an increasingly common experience for US scientists—an experience that, as we have seen, not everyone regarded favorably. In many of the critiques about refereeing at journals, we see scientists expressing frustration over what they saw as an editorial abdication of duty. Some authors felt that editors should be the ones evaluating the papers and saw the increasing US reliance on refereeing as a sign that editors were not doing their jobs. The division of responsibility between referees, editors, and funding body employees would become a subject of recurring controversy in the 1970s, as we shall see in chapters 5 and 6.

However, for the most part, it appears that US science adapted to increased refereeing with relative ease. Editors found referee opinions useful for managing their workload and collecting advice on fields unfamiliar to them; federal funding bodies saw scientific advice as a useful source of expert knowledge; and scientists proved generally willing to provide their opinions and (with exceptions) willing to hear others' opinions on their work. It is worth noting once more that, in this period, the default decision at most funding bodies and journals was still "accept," which likely helped ease the sting of critical comments. By 1965 refereeing was an expected part of grant applications and journal submissions for US scientists.

As refereeing expanded and became more commonplace, another group of researchers took note: sociologists and philosophers in the field of science studies. It was these scholars who first articulated a vision of refereeing as not just a means of collecting advice but as a system that underpinned trust and credibility in the sciences.

4

FROM REFEREEING TO PEER REVIEW

Readers may have noticed that in the first three chapters of this book, I have exclusively used the term "refereeing" rather than "peer review." There is a simple explanation for my stubbornness: Before the 1960s, the term "peer review" did not exist. George Gabriel Stokes would have been quite baffled if someone asked him whether the *Philosophical Transactions* was peer-reviewed. Phil Abelson would have been similarly confused about the question in 1962 when he began editing *Science*. Historians like to stick to actors' categories—in other words, to use the terms and phrases that would have made sense to the people we're writing about—hence my insistence on the term "refereeing" so far. By the end of this chapter, that will change. We will finally arrive in the late 1960s and the early 1970s, when "peer review" became an actors' category among scientists.

As we saw in the last chapter, by the 1960s refereeing was not a universal practice at either journals or funding bodies, although US journals and funding bodies were becoming more likely to use it. However, researchers in the philosophy and sociology of science were beginning to take an interest in refereeing at scientific journals. Some of those scholars, most notably the Columbia sociologists Robert Merton and Harriet Zuckerman, saw refereeing not just as an optional bureaucratic process. They saw refereeing as a mechanism that ensured the reliability of scientific knowledge claims and fostered trust between scientists. It was sociologists and philosophers who first began describing refereeing not just as an occasional part of science but as one of its most important features—and as a practice that only worked because it was scientific experts who were doing the refereeing.

Around the same time that scholars like Merton and Zuckerman were promoting refereeing as a central part of science, an important linguistic shift was taking place: Increasingly, Americans were calling refereeing "peer review," particularly in the context of grants. The use and spread of the term "peer review" is more than a linguistic curiosity. The term established clear limits on who was allowed to participate in the practice of reviewing a grant proposal. To be "peer reviewed," a grant had to be evaluated by scientific experts, the intellectual peers of the person who had written the proposal. The term "peer review" thus echoed many of the claims that sociologists and philosophers had been making about the importance of expert evaluation in science.

REFEREEING ACCORDING TO SOCIOLOGISTS AND PHILOSOPHERS OF SCIENCE

The origins of sociology and philosophy of science are generally traced to the early twentieth century in Europe and to figures such as J. D. Bernal, J.B.S. Haldane, Boris Hessen, and Nikolai Bukharin. These writers, many of whom were also scientists, drew on Marxist philosophy to study both scientific theories and the structure and function of scientific institutions.[1] Other early figures, such as the Hungarian physical chemist turned social scientist Michael Polanyi, rejected the Marxism of early sociology of science but joined in the project of using sociology, history, and philosophy to analyze how—or if—science could arrive at reliable knowledge claims.[2] By the early 1960s, sociology of science and philosophy of science were established subdisciplines in their respective fields, although the institutionalization of "science studies" as its own distinct field would not come until later.

One of the most receptive audiences for work in the philosophy and sociology of science in the 1960s and 1970s was scientists themselves. During the early decades of the Cold War, members of the natural sciences community were deeply interested in and engaged with the work of key figures such as Polanyi, Merton, Thomas Kuhn, Karl Popper, John Ziman, and Derek de Solla Price. Reviews of books by sociologists, historians, and philosophers of science regularly appeared in scientist-centered publications such as *Science*, *Nature*, and *Physics Today*.[3] Kuhn's 1962 book *The Structure of Scientific Revolutions*, which described the history of science as a series of "revolutions" in which an older scientific "paradigm" was replaced with a newer one, was especially widely reviewed. Scientific publications also welcomed articles written by sociologists and philosophers of science, reflecting their readers' interest in that work.[4] A number of the most influential writers in those fields had been trained as scientists originally, including Ziman and de Solla Price, further bolstering the impression that these thinkers took science seriously and truly understood its inner workings. Scientists even began describing their own work using terms from the philosophy and sociology of science. Geologists in the 1960s, for instance, explicitly cast their field's adoption of the "continental drift" model as a case of a Kuhnian "revolution," in which the new continental drift "paradigm" was pushing out the older vision of the earth's crust as static and unmoving.[5] Natural scientists embraced Merton's work on the norms of science (which we will discuss shortly), enthusiastically agreeing that features like disinterestedness and organized skepticism were the guiding principles of scientific work.[6] In 1982, the science journalists Nicholas Wade and William Broad would actually bemoan the influence of scholars like Kuhn and Merton, writing that they had given scientists an unreasonably lofty self-image.[7]

Scientific publishing and grants were not initially a strong area of interest among early figures in the sociology or philosophy of science. However, there were occasional references to refereeing in pre–Cold War works. Polanyi, for instance, argued that multiple referees were likely to agree on the merit of a scientific paper, and used the agreement between referees as evidence for the reliability of scientific training.[8] In his influential 1939 book *The Social Function of Science*, Bernal, a Marxist crystallographer who became known for his writings on the relationship between science and society, expressed frustration at the system of scientific communication. There were too many journals, he said, and it was virtually impossible for scientists to find relevant articles in their field. He suggested that subscriptions to journals should be replaced by the circulation of articles. He envisioned a system in which a scientist "sends his paper in to the publication office; it is then submitted to an editorial board of referees as at present and, if passed, amended or unamended, is reproduced photographically."[9] The articles would then be mailed to libraries and researchers who had requested work in that field. Bernal's comments indicate that in his mind, editorial board review and refereeing were one and the same; he saw value in expert review but envisioned a limited editorial board rather than a broad swath of external referees as the group performing it.

By the 1960s, a larger group of sociologists and philosophers of science were turning their attention to scientific publishing. Several of those writers viewed refereeing with skepticism. The economist Gordon Tullock, for instance, believed refereeing was an inefficient system that was unlikely to find the best expert advice. In his 1966 book *The Organization of Inquiry*, Tullock argued that choosing the best papers for publication was "of the utmost importance" to the proper functioning of both the natural and social scientists, and therefore, the best scientific talent should be recruited to editorial positions. In Tullock's mind, sending a paper to an external reader for advice was illogical: "If the reader is really more competent [than the editor], then he should be editor; if he is less competent, he should not be given the deciding voice." The ideal system, argued Tullock, was one in which each journal had an eminent editorial board with wide-ranging expertise and all decisions were made by the editorial team.[10] The American sociologist of science Warren O. Hagstrom described scientific papers as gifts to a scientist's peers, given because scientists were motivated by "the desire to obtain social recognition."[11] Hagstrom downplayed the significance of referees in publishing because of the low rejection rate at most scientific journals and because many journals were unrefereed. He also did not see increased reliance on referees as a good path for science: "Delegating considerable power to a few authorities obviously infringes on the norms of independence in science," he argued.[12] For Hagstrom, the most important judgment of a scientist's work took place after publication, not before.

Other scholars argued that refereeing inherently favored established ideas and did little to weed out poor papers. In a 1963 paper for the sociology of science journal *Minerva*, for example, the physicist Alvin

Weinberg said that the recent "information explosion" in scientific journals meant that scientific papers were not read carefully: "Nonsense is not so generally recognised as such, and the standards of self-criticism, which are so necessary if the scientific literature is to serve as the arbiter of scientific taste, are inevitably looser than they once were." Furthermore, said Weinberg, the refereeing and editorial system prevented meaningful criticism: "A referee of a scientific paper asks whether the paper conforms to the rules of the scientific community to which both referee and author belong, not whether the rules themselves are valid. So to speak, the editors and authors . . . are all tainted with the same poison."[13]

But other scholars ascribed far more importance to the work of a referee, particularly in the mid-1960s as some scientists began to push for wider circulation of preprints and drafts as an alternative to published journals.[14] Opponents of a preprint-centered system argued that there was no quality control over preprints and that only papers selected for publication in a journal should be considered legitimate contributions to scientific knowledge. De Solla Price, a British expat and former physicist in the Yale Department of the History of Science and Medicine, became known as the founder of a field called "scientometrics," which used numerical data about scientific papers to track changes in science. In a 1964 paper for *Science*, de Solla Price argued that scientific publishing ought "to be considered a privilege consequent upon the finding of something which people may need to read, rather than as a duty consequent upon the spending of time and money" and said that to join the scientific literature, "a paper has to run the gauntlet of editors and referees."[15] In his 1968 book *Public Knowledge*, the British physicist turned philosopher of science John Ziman argued that in science, "the 'literature' of a subject is quite as important as the research work that it embodies"[16] and said that refereeing and editing were the reasons why the scientific community should prefer selective academic journals to preprints or a journal that published every submission. Referees and editors, said Ziman, bestowed "the *imprimatur* of scientific authenticity" because readers could trust that experts had reviewed the article and deemed it worthy of publication.[17] "An attack on the function of the referee, whether by the pseudo-publication of 'preprints' or by the actual publication of unrefereed 'letters' journals and such like rubbish, is a blow at the roots of Science itself," he declared.[18] The referee was, and ought to be, "the lynchpin about which the whole business of Science is pivoted."[19]

The most influential sociological work on publishing and refereeing, however, was that of Robert Merton and Harriet Zuckerman. Born Meyer R. Schkolnick, the son of Jewish immigrants in south Philadelphia eventually adopted the name "Robert K. Merton" as part of a career as an amateur magician.[20] After a successful undergraduate career at Philadelphia's Temple University, Merton decided to study in the new department of sociology at Harvard University, where he defended his PhD in 1935. By 1942, Merton had outlined his famous "Mertonian norms" of science in two now-classic papers, arguing that science was held together by the shared behavioral norms of universalism, communism, organized skepticism, and disinterestedness.[21] Universalism dictated that science was a field and practice that anyone could participate in; communism required that scientists publicly share their discoveries with their peers; organized skepticism required that scientists assess their own and others' scientific work with a rigorous, skeptical eye before accepting its conclusions; disinterestedness required scientists to put the good of science before their own professional goals.

Although some of Merton's mentors encouraged Harvard to hire him as a faculty member, he was not offered a position at his alma mater, at least in part because of his Jewish heritage. Instead, in 1941, Merton was hired at Columbia, where he would go on to become one of the most influential American social scientists of the 1950s and 1960s. At Columbia, Merton began collaborating with Professor Paul Lazarsfeld and the graduate student Patricia Kendall, with whom he developed the first sociological focus groups. Merton's postwar sociological interests were wide-ranging. Fueled and funded by a new interest in applied sociological research, Merton studied business organizations and bureaucracy, the sociology of housing, racial integration, and mass communication. He also retained his interest in, and influence over, the field of sociology of science. His status in the field is clearly reflected by the fact that in the early 1970s, the NSF called on Merton personally to ask for his assistance in developing indicators to determine how they were doing in selecting and funding proposals.[22]

Under Merton's leadership Columbia became a major training ground for PhD students. One of those students was Merton's future collaborator Harriet Zuckerman, who had begun studying sociology as an undergraduate at Vassar College in the 1950s. Zuckerman initially chose the sociology department at the University of Chicago for her graduate work, in large part because her new husband wanted to move to Chicago. However, as she would later tell the story, both the marriage and the department began falling apart shortly after she arrived. When Lazarsfeld visited Chicago, she said she was "absolutely captivated" by his ideas and made the decision to transfer to Columbia in 1959.[23]

By the time Zuckerman arrived at Columbia, one of Merton's many research areas was scientific communication and publishing. As Alex Csiszar has observed, studying publishing was a way for Merton to try to observe his famous norms in action; by analyzing which papers made it into print, how they were selected, and how they were received by their readers, Merton could evaluate how the scientific community distributed professional rewards for scientific work.[24] In two major papers, one in 1957 and another in 1968, Merton analyzed how various professional rewards such as credit for theories, major prizes like the Nobel, and desirable jobs at prestigious institutions were distributed in science and whether that distribution encouraged the creation of new scientific knowledge.[25] Merton argued that the desire for recognition from their peers was the key factor that motivated scientists to create new scientific knowledge, and he considered scientific journals one of the most significant institutions shaping scientists' behavior. As Merton put it in an article for the general-audience magazine *American Scientist* in 1969, the first scientific journals "motivate[d] men of science to replace the value set upon secrecy with the value placed upon the open disclosure of the knowledge they had created."[26]

One of the major data sources Merton utilized for the 1968 paper was Zuckerman's interviews of Nobel laureates. Merton would later say that Zuckerman should have been credited as a coauthor and expressed regret that he had failed to do so.[27] Zuckerman, who defended her PhD in 1965 and became an assistant professor at Columbia shortly thereafter, would soon become one of Merton's most frequent—and this time, credited—collaborators.[28]

"PATTERNS OF EVALUATION IN SCIENCE" (1971)

Shortly after Zuckerman defended her PhD, she and Merton began collaborating on a fine-grained study of refereeing at *Physical Review*. In 1971 the pair published an article in the science studies journal *Minerva* that analyzed the referee system in the natural sciences. The article would prove to be the single most influential paper on refereeing ever published, in any field. Zuckerman and Merton approached their analysis of refereeing along four paths. First, they gave an account of what they saw as the history of the refereeing system. Second, they analyzed patterns of rejection rates at scholarly journals. Third, they analyzed the archives of *Physical Review* "to identify and analyse patterns of decision by editors and referees." Their concluding section focused on "the significance of the referee system for individual scientists, scientific communication and the development of science."[29]

The historical section of the paper built on Merton's earlier work on seventeenth-century science, drawing particularly on his knowledge of the correspondence of the chemist Robert Boyle. The paper largely credited the *Philosophical Transactions* editor Henry Oldenburg with inventing prepublication review by experts, and it argued that one of the factors drawing natural philosophers to publish in the *Philosophical Transactions* was a desire for "competent appraisals of their work by other authoritative scientists."[30] Their historical section made a brief mention of article review at the *Journal des Scavans* before concluding with the claim that "almost from their beginning, then, the scientific journals were developing modes of refereeing for the express purpose of controlling the quality of what they put in print."[31]

Readers have likely noticed that Zuckerman and Merton's version of the first years of refereeing differs substantially from the story I told in chapter 1. Historians of scientific publishing now agree that the two sociologists did not quite get the history right in their influential paper. There is a tendency in their account to interpret similarities as sameness, to look back at the past and conclude that anything that looks familiar must be the origins of the practices we know today. But we know from reactions to Whewell's proposal in the 1830s that refereeing did *not* take a firm hold in scientific publishing in the 1660s and persist unchanged through the twentieth century. Whewell's contemporaries considered the idea of having experts review papers prior to publication novel. The idea that "quality control" was the primary purpose of early refereeing is also debatable. For the purposes of this chapter, however, Zuckerman and Merton's historical accuracy is less important than their influence on their contemporaries—which, as we shall see, was substantial.

The next section of Zuckerman and Merton's article compared rejection rates for 83 journals in the natural sciences, social sciences, and humanities. They found that humanities journals rejected the largest number of papers—the three history journals they consulted, for instance, rejected an average of 90 percent of papers—while the natural sciences journals, on average, rejected under a third of submissions. The social sciences fell somewhere in the middle, with rejection rates ranging from 78 percent in sociology to 48 percent in anthropology. Zuckerman and Merton suggested that journals with high rejection rates received a large number of submissions that "fail by a wide margin to measure up even to minimum standards of scholarship;"[32] consequently, many papers received an editorial rejection and were never seen by a referee. At low-rejection journals, however, editors assumed "that a manuscript is publishable until clearly proved otherwise."[33]

Zuckerman and Merton then turned their attention to a case study of *Physical Review*. To justify their selection, they pointed out that *Physical Review* had become central to physics publishing, and not just in the United States: "In such leading journals as the Italian *Nuovo Cimento*, the *Russian Journal of Experimental Physics* (*JETP*), and the *Proceedings* of the Physical Society of London, the *Review* is cited far more often than these journals themselves."[34] Zuckerman and Merton drew on "correspondence between authors, editors and referees, records of decisions made by the editors, the allocation of manuscripts to referees, their evaluations and the final disposition of the paper."[35] The analysis was applied to *Physical Review* between 1948 and 1956, prior to the split into multiple journals described in chapter 3.

To cope with a large amount of material—14,512 submissions during the study period—Zuckerman and Merton chose to focus on a sample of physicists who submitted single-authored papers, sorted by a metric that the pair called "institutionalized standing." They separated authors of submitted manuscripts into three categories. The first, highest-status category included physicists who "had received at least one of the ten most respected awards in physics (such as the Nobel prize, membership in the Royal Society and in the National Academy of Sciences)."[36] Just fifty-five physicists in category 1 had submitted single-authored papers to *Physical Review* in the study period. The second tier applied to 343 potential contributors who did not have a major prize but were included in the "Archive of Contemporary Physicists," a sort of "who's who" of physics maintained by AIP. Physicists who did not qualify for the first two tiers were assigned to the third tier; since that tier was so large, Zuckerman and Merton confined their analysis to a random sampling of 659 aspiring contributors. They applied the same tiers to the 354 referees whose work they analyzed, writing that 12 percent of referees were in the first rank, 35 percent in the second, and 53 percent in the third. Physicists in the top rank submitted an average of 4.09 papers to *Physical Review* over the study period; second rank averaged 3.46; third rank averaged 2.02.

Strikingly, a full 78 percent of papers from rank 1 physicists were accepted by editors alone, without being sent to a referee. Rank 2 physicists had a 70 percent chance of editorial acceptance without refereeing. Fifty-eight percent of papers by rank 3 physicists were accepted by editors without refereeing. One important consequence of the likelier desk accept was faster publication; 89 percent of papers from top-rank physicists made it to print in under four months, versus 70 percent for third-rank physicists. Rejections without a referee were comparatively rare, with only 12 percent of papers overall rejected without being sent to referees.[37] The remaining 23 percent of manuscripts, which Zuckerman and Merton deemed "problematic manuscripts," were sent to outside referees.[38] The majority of papers sent to referees were ultimately accepted after major revisions. Just 8 percent of papers suffered the fate of being rejected after receiving referee opinions.

Zuckerman and Merton analyzed the recommendations from referees to see if status was exerting a significant influence over referee reports. They concluded, however, that there was no evidence of significant referee status bias. "Referees of each rank accept the same proportion of papers by authors from every stratum," they wrote. "As it happens, the highest ranking referees accept somewhat smaller proportions of papers than their fellow referees but, again, this they do uniformly for authors of every rank. There is, in short, no preferential pattern."[39] The pair felt confident that *Physical Review* referees were basing their judgments on scientific merit rather than on status or reputation.

From this data, Zuckerman and Merton reached a striking conclusion: that "the referee system provides for a warranted faith that archives of science can generally be relied upon."[40] In other words, refereeing was the reason that readers could open a scientific journal and trust that the claims within it were reliable. Authors took more care preparing their papers knowing that a manuscript might be sent to a referee, referees weeded out obviously poor or flawed papers, and scientists reading in a field outside their own specialization could rest easy knowing that an expert in a paper's field had endorsed the manuscript before it was published. "The system of monitoring scientific work before it enters into the archives of science means that much of the time scientists can build upon the work of others with a degree of warranted confidence," the pair wrote in the paper's final lines. "It is in this sense that the structure of authority in science, in which the referee system occupies a central place, provides an institutional basis for the comparative reliability and cumulation of knowledge."[41] The referee system was the key to securing trust in not just the scientific literature, but in science itself.

REFEREEING IN THEORY VERSUS PRACTICE

It would be hard to overstate the influence "Patterns of Evaluation in Science" had on perceptions of refereeing. The interest in the paper from both fellow sociologists and researchers in the natural sciences was immediate. *Physics Today*, the flagship magazine of AIP, published an abridged version of the study in their July 1971 issue.[42] "Physicists have a right to feel proud," the *Physics Today* editor Harold Davis wrote in his editorial for the month. "Harriet Zuckerman and Robert Merton, two eminent sociologists of science, have conducted a thorough assessment of the refereeing system that serves *Physical Review*. They report that the system is doing an excellent job."[43] By the time the article came out, its data on *Physical Review* refereeing was over fifteen years old, but that did not create much of a problem for their claims; in that time, *Physical Review* had grown even more reliant on referees, entirely eschewing the practice of "editorial acceptance."[44] In a 1986 interview, Zuckerman described the *Physical Review* study as one of her most significant achievements at Columbia, in part due to its continued citations in discussions of modern peer review.[45] One way to trace the impact of the article is by looking at the way the Zuckerman-Merton origin story for peer review—that it was invented by Henry Oldenburg and had been a more or less expected part of science ever since—became a fixture of the scholarly literature about peer review. Time and again over the successive decades, sociologists, physicists, physicians, policy scholars, and biologists cited Zuckerman and Merton's work to back up the claim that refereeing had been part of scientific publishing since the Scientific Revolution.[46]

However, a broader look at discussions of refereeing and editing in the mid- to late 1960s suggests that philosophers and sociologists of science were promoting a vision of refereeing that was out of step with the way refereeing was actually practiced. Ziman's description of the referee as the "lynchpin" of science—a description Zuckerman and Merton quoted in their opening paragraph—was evocative but not quite accurate at the time he wrote it. It is worth emphasizing that Zuckerman and Merton themselves found that in the years of their study, over 50 percent of the accepted papers at *Physical Review* never saw a referee at all. Although *Physical Review* had moved away from editorial acceptances by the time the *Minerva* paper was published in 1971, the high-profile spinoff *Physical Review Letters* did not require two referee reports until 1972. Many other high-profile journals still accepted some papers without any reports, especially outside the US, and publication in those journals was not yet seen as inferior to publication in refereed journals.

Furthermore, as we saw in chapter 3, the idea that referees might improve the quality of a journal was relatively new in the mid-1960s. Scientists and editors still tended to discuss refereeing in practical terms, as an aid for editors rather than a crucial aspect of science itself. For instance, in 1965 retired chemist Bernard K. Forscher proposed a set of "Rules for Referees" in an opinion piece for *Science*. Refereeing, said Forscher, had "great merit" due to the fact that "one cannot expect the editor of a journal to have knowledge of the background and methods" of every paper submitted to his journal.[47] But, said Forscher, the referee should never have too much power. "The decision to accept or reject is the duty of the editor, not of the referee,"[48] he wrote, adding that "the referee's report is specifically for the guidance of the editor."[49] Similarly, in a 1976 editorial for *Physical Review Letters*, the editors James Krumhansl, George Trigg, and Gene Wells emphasized that "referees are expected to make their reports to the editors, not to the authors" and "it is our job to decide what action to take."[50] The *Physical Review* editor Pasternack spoke passionately about the value of refereeing in a 1966 *Physics Today* essay, but he cast its benefits as practical ones, such as making articles easier to read.[51]

Refereed articles were also only beginning to receive special status when universities considered tenure and promotion cases in the sciences. In the mid-1970s, the vice-president for academic affairs at the University of Maryland College Park undertook a review of tenure standards across campus. As part of that review, each department submitted a report about its requirements for tenure, and several departments in the natural and social sciences specifically mentioned refereed articles as the appropriate basis for tenure decisions. The psychology department's guidelines stated that publications had to "meet the quality standards of the better refereed journals"[52]; the department of human ecology specified that "publications in refereed journals and in books" were the appropriate evidence of scholarly activity.[53] The department of physics and astronomy instructed its tenure committee to collect copies of "all papers accepted for publication in refereed journals."[54] However, other departments in the 1970s did not specifically mention refereed articles in their promotion criteria, including the departments of mathematics, zoology, microbiology, geology, and sociology; all research articles seemed to count equally according to those departments' guidelines.

This gap between discourse and practice highlights how novel "Patterns of Evaluation in Science" was in articulating a vision of refereeing as a process that was central to the creation of scientific knowledge. For Zuckerman and Merton, refereeing wasn't just a way to lessen the burdens on editors, as it had been for *Science*, or a way to improve the quality of papers, as it had been described to AIP journal referees in the 1960s. It was key to "the reliability and cumulation of knowledge." And Zuckerman and Merton made this argument knowing full well that many journals still did not use referees. They acknowledged that fact in their paper but argued that their conclusions were still correct because refereed journals raised the standard of quality for the entire literature—a leap in logic that highlighted the gap between their epistemological vision and the actual working reality of scientific journals.

Nevertheless, Zuckerman and Merton were confident in their view that refereeing occupied a "central place" in "the structure of authority in science." Without referees, scientific publishing, and therefore science itself, simply would not work. And the key to a functioning referee system was the opinions of experts in a paper's field prior to publication—the scientific peers of the paper's authors.

Which brings us, finally, to the term "peer review."

THE RISE OF "PEER REVIEW"

Zuckerman and Merton themselves did not employ the term "peer review" in "Patterns of Evaluation." Instead, they consistently used "refereeing," and would not have been seen as strange for making that choice. The term "peer review" appears to have first been used not in reference to scientific research but to describe review committees in the medical community. In the 1960s and early 1970s, for example, major American newspapers usually used "peer review" to refer to reviews of medical practices for compliance with Medicare and Medicaid.[55] At the NIH in the 1950s and 1960s, the term "peer" frequently appeared in discussions of the NIH's reviewing procedures. NIH staff members and applicants alike praised the NIH's system of having study sections of scientific experts, the scientific "peers" of applicants, who made recommendations on the scientific merit of proposals. However, the 1965 Wooldridge Committee report on the NIH, a sweeping analysis of the NIH's bureaucratic procedures that included extensive discussion of the NIH's proposal reviewing mechanisms, did not use the term "peer review" at all, instead speaking of "peer judgments."[56]

One useful way to track how the phrase "peer review" spread to grant review in the English-speaking scientific community is to look at *Science* and *Nature*, two prominent scientific weeklies—one American, one British—that contained news and commentary. *Science* first used the term "peer review" in a 1965 article about the Wooldridge Committee Report.

Joseph D. Cooper explained that as part of an evaluation of the NIH's review process, "project grantees . . . were asked to make assessments of the output of a system of project approval in which they were intimately involved both as grantees and as members of peer review groups."[57] However, *Science* used the term only infrequently until the 1970s, and then mostly in reference to review procedures specific to medical journals and the NIH.[58] Across the Atlantic Ocean, the British *Nature* did not use the term at all until 1971 and, for most of the 1970s, used it only to refer to American grant or institutional review in the biomedical sciences.[59] One *Nature* article in 1975 described "the so-called peer review system" specifically as a "method for judging the relative merits of competing grant proposals."[60]

In the medical literature, the *New England Journal of Medicine* (*NEJM*) first used the term "peer review" in 1969 to refer to the procedures for ensuring that hospitals and physicians were abiding by the regulations of Medicaid federal funding programs:

> Although state health authorities have been concerned with standard formulation and facility licensing and regulation for many years, few state programs can boast of the establishment of effective monitoring or "peer review" methods to assure the adequacy of institutional and professional services under Medicaid.[61]

Similarly, the term was first used in the *Journal of the American Medical Association* (*JAMA*) in 1970 to describe physicians' assessments of each others' practices under the Professional Standards Review Organization (PSRO) associated with Medicaid; this continued to be *JAMA*'s most frequent use of the phrase through the 1970s.[62] "Peer review means different things to different people," the physician Irvine H. Page explained in a 1973 editorial for *JAMA*. "To most American physicians it means PSRO, to the British House of Lords it means Peers examining other Peers for moral turpitude, and to the scientific community, it means Study Sections and Councils that determine a grantee's financial and possibly research future."[63] Notably, journal refereeing was *not* one of the definitions Page offered, and *JAMA* would not use the term "peer review" to refer to journal refereeing until 1981.[64]

Contemporary commentary from the 1970s makes it clear that outside the US, the term "peer review" was seen as an Americanism. In

December 1972, the British scientific journal *Nature* ran a letter in its "Correspondence" column from the zoologist Solomon Zuckerman, who had served as the British Government's chief scientific advisor from 1964 to 1971. Zuckerman was writing to defend the advice he had given the government about grants for cancer research, in which he had said that scientists might bring personal biases into their assessment of grant applications.[65] This was not, he said, an argument against using scientific experts to help review grants. In fact, he believed there was "no sensible alternative" to "what the Americans call 'peer review'" or to "the review process whereby a scientific paper is judged worthy of publication."[66] Significantly, as of 1972, Zuckerman thought of "peer review" as not only an Americanism but one that was specifically about grant review and not journals.

Franz J. Ingelfinger, the influential editor of the *New England Journal of Medicine,* was one of the first people to refer to journal refereeing as "peer review" in print. In 1968, Ingelfinger was invited to write an opinion piece for *The Lancet* on the purpose of the medical journal. He wrote that "to be accepted the article must pass peer review, and ideally either must pass with high honours or must undergo appropriate revision."[67] The term would not appear again in *The Lancet* until the 1970s. Ingelfinger was also the first to use the term "peer review" for journal refereeing in *Science,* which he did in a 1970 article on the current state of medical literature.[68] In a 1970 article on *NEJM*'s review process, a member of the *NEJM* editorial staff (likely Ingelfinger) wrote that the journal tried to maintain quality and integrity through "a 'peer review' of submitted articles whose findings are weighed in final judgment by editors and their advisers."[69] The biologist Thomas Jukes, an Englishman at the University of California-Berkeley and a regular *Nature* columnist, was the first *Nature* author to call journal refereeing "peer review," in 1977.[70] By the early 1980s, most observers used the term "peer review" to apply to review at both journals and funding bodies.

The shift from "refereeing" to "peer review" is not merely a linguistic curiosity. Calling external refereeing at journals or funding bodies "peer review" made it explicit that the evaluation of a paper or grant proposal could only be done by experts—the peers of the person who had submitted their work. The new term established a narrow range of acceptable reviewers and deemed those without a scientific background unqualified to evaluate the work in question. That was very much in line with the vision of refereeing that Zuckerman and Merton had promoted in their work, in which refereeing was crucial for scientific knowledge-making because it meant a piece of scientific work had been evaluated by experts. It is also significant that the term seems to have first been created in the area of grant review but expanded to journals beginning in the late 1970s, reflecting an increasing conflation of two very different types of refereeing.

CONCLUSION

In the first three chapters of this book, the motivations we've seen for adopting referee systems were almost entirely practical ones. Editors needed help with their workloads, new federal funding agencies used scientific experts to establish their own legitimacy, and learned societies wanted to protect their reputations. The idea that refereeing serves an epistemological function—that it has something to do with how we know whether a claim is true or reliable—does not seem to have originated in the natural sciences at all. That idea, as I've worked to show in this chapter, seems to have come from the philosophy and sociology of science.

However, there was a noticeable gap between the scholarly discourse about refereeing and the way refereeing worked in practice. Discussions of refereeing from sociologists and philosophers during this period were often aspirational rather than descriptive; they reflected a high-minded theory of refereeing that was not necessarily matched by the way refereeing actually took place in the sciences. Refereeing was growing increasingly common in the US in the years leading up to the Zuckerman-Merton paper, but the reasons for that were practical rather than epistemological. Refereeing spread because it made running journals and funding bodies easier, not because scientists believed refereeing made the final outcomes of those things more reliable or trustworthy.

And yet, the idea that refereeing improved the quality of science and gave stakeholders a reason to trust scientific claims took hold with remarkable speed. As a changing political situation in the United States led to increasingly close scrutiny of science funding in the 1970s, scientists sought to justify their role as advice-givers at federal funding bodies—and would adopt language from the sociology and philosophy of science about the importance of "peer review" to do so. The next two chapters will focus on two controversies that challenged peer review's status in the US: one at the NIH and one at the NSF. Both controversies ultimately served to cement peer review's growing importance to science in the US.

5

"ABOLISH THE PEER REVIEW PROCESS"

THE NATIONAL INSTITUTES OF HEALTH VERSUS THE OFFICE OF MANAGEMENT AND BUDGET

In the spring of 1973, the US Office of Management and Budget (OMB)—a recently created arm of the executive branch tasked with overseeing the financial aspects of the president's agenda—released a report on grant review at the National Institutes of Health (NIH). President Richard Nixon had made headlines with his "War on Cancer," an ambitious program that sought to make major progress in cancer diagnosis and treatment. The NIH was supposed to be a significant source of funding for the research that would achieve the Nixon administration's goals.[1] The OMB, however, had serious reservations about the way the NIH was awarding its grant money.

Specifically, the unnamed authors of the OMB report found the NIH's "peer review" system baffling. The natural method of awarding federal money, they felt, would have been for NIH administrators to decide which applications to approve. The NIH's expert reviewers were not federal employees and were, thus, entirely unaccountable to the public and to the president. Furthermore, scientific experts were probably making decisions on proposals put forward by professional contacts, colleagues, or even collaborators, which made the system rife with opportunity for conflicts of interest. The OMB report listed several suggestions for how the NIH might improve its grant review process, but it led with the one that the OMB report authors seemed to think the most desirable: "Abolish the Peer Review Process, Relying Completely on Administrative Review."[2]

This chapter argues that the OMB controversy, along with a congressional hearing about research ethics that same year, shaped the language US scientists used to discuss peer review with US legislators and policymakers. The War on Cancer prompted the Nixon administration to pay closer attention to the NIH and to try to take a more active role in shaping NIH grant-making and spending. That scrutiny led to conflicts between the Nixon administration, on the one hand, and the NIH and the research community on the other. The OMB questioned whether scientists were really the best people to advise a massive federal body like the NIH on how to spend its money. In response, US scientists began characterizing peer review as the best system for awarding grants, a system that would not only yield the most qualified advice but would prevent the government from pursuing harmful or unethical projects. In arguing that review by a scientist's peers was the best way to decide which projects to fund, scientists were explicitly seeking to carve out and protect a role for themselves in federal science funding decisions.

NIH REVIEW IN THE 1960s AND 1970s

An overview of how NIH review worked in the middle decades of the Cold War will be useful for understanding the OMB's criticisms of that process. By the early 1970s, the NIH consistently described their review process using the term "peer review"—and indeed, as we saw in chapter 4, seems to have been the first funding body to do so. Another term that appears frequently in NIH communications from this period is "dual review," a term meant to call attention to the fact that there were multiple steps to the review of an NIH proposal. Although the terminology had evolved, the system overall was very similar to the one used in the NIH's first years (see chapter 3). The first round of proposal review was performed by a "study section," a formally appointed group of experts in a specific field like "Allergy and Immunology," "Endocrinology," "Genetics," or "Toxicology." As of 1965 there were roughly fifty study sections.[3] The study sections awarded a grade of "approve" or "disapprove" to each proposal based on the criterion of "scientific merit." From there, the "approved" applications would go to the NIH advisory councils, mixed groups of scientists and laymen who decided which of the approved proposals would ultimately be funded. Final decisions lay in the hands of the directors of NIH member institutes such as the National Eye Institute or the National Cancer Institute.

Not every proposal evaluated at the NIH went through exactly the same process. In addition to its research grants, the NIH awarded contracts to groups working toward specific goals or on specific scientific questions. Review and approval procedures for contracts seems to have been more eclectic than the highly standardized grant review process. Outside experts were often consulted, but not systematically, and not by the study section mechanism. It was study section review of grants that tended to receive the lion's share of attention from both the NIH itself and the rest of the government, however, and the "peer review" system was a point of intense pride for NIH leaders. In a 1972 report to the secretary of health, education, and welfare, the NIH administrator A. A. Kaufman explained NIH "peer review" with the following description.

> The peer review system is designed to assure the high quality of biomedical research supported by the Federal government, through careful appraisal of the quality of applications for support. It is based on the premise that the individuals best qualified to pass on the scientific and technical merits of a research proposal are those scientists most knowledgeable in the field of the proposal.[4]

As we saw in chapter 3, the NIH's study section system was different in important ways from the procedures used at scientific journals that employed refereeing. The NIH did not seek out the most qualified reviewers for each individual proposal; instead, it appointed twelve to fifteen experts to each study section, and those experts served four-year terms. Everyone who evaluated grant applications was, therefore, in some way formally affiliated with the NIH, even if they did not draw a salary from the NIH. The system relied on the group's collective experience to be able to evaluate all proposals, rather than trying to match each grant proposal with reviewers who had expertise in its subject. The NIH's system was, thus, more akin to employing a rotating editorial board to review all papers than to referee systems that sent each paper to two or three independent reviewers whose expertise was a close match with the paper's field. Furthermore, the system of evaluating grant applications in person during a group meeting was unlike the refereeing-by-mail system that most journals used.

The NIH's system was also distinct from most systems used at other federal agencies that awarded scientific grants. Most federal agencies consulted outside referees by mail or by phone, a method more like the referee systems at journals. In one NIH report from 1973, NIH employees cataloged grant review processes at ten government agencies: the Atomic Energy Commission (AEC), the Department of Agriculture, the Office of Naval Research (ONR), the US Air Force, the US Army, the Environmental Protection Agency (EPA), the Food and Drug Administration, NASA, the NSF, and the Veterans Administration (VA). There were four broad categories of proposal review type: "outside readers," in which the agency solicits comments on proposals by external experts via mail; "panel review," in which panels of experts are assembled in person to evaluate grant applications; "ad hoc consulting," in which external experts are occasionally consulted but many proposals do not receive outside feedback; and finally, "internal review," in which review of proposals was done by permanent employees of the grant agency.

Notably, the NIH had by far the largest biomedical research budget of any of these organizations. The Department of Agriculture awarded around $4.5 million in grants; the FDA funded $3.2 million in projects; the ONR was spending under $30 million on contracts and grants for biomedical and psychological research. The largest non-NIH research budgets belonged to the VA ($70 million) and the NSF ($60 million). The NIH, however, had awarded $1.48 billion for research in FY 1972—a staggering twenty-one times larger than the next-largest budget.[5]

At all ten agencies, final funding decisions lay in the hands of agency employees. However, no other agencies took the NIH's approach of relying exclusively on panel review to advise on the scientific merit of a grant proposal. Three of the ten—the AEC, the Department of Agriculture, and the EPA—used outside readers exclusively, relying on mailed hard copies of proposals to collect opinions. The NSF, NASA, and the FDA used a combination of outside readers and panel review.[6] The ONR and the US Army each used a combination of in-house review and outside readers, although the report noted that the Army supported "very few research grants," and at the ONR, "the review process is not bound by rigid policies or procedural requirements." The Veterans Administration review process employed both review panels and staff site visits to hospitals that received or might receive grant money. The US Air Force used internal review alone to evaluate scientific or technical merit.[7]

The NIH was well aware that its budget was much larger than that of comparable federal programs and that its procedures for awarding money were unusual. In the decades after its foundation, NIH leaders went to great lengths to explain and defend their system to new presidential administrations and political appointees. Between 1947 and 1971, there were eight internal studies of NIH peer review, each one meant to explain the process and evaluate its effectiveness.[8] The most significant was the 1965 report *Biomedical Science and its Administration: A Study of the National Institutes of Health,* usually called the "Wooldridge Committee report" after its chair, engineer Dean Wooldridge. That report found that 70 percent of the NIH's annual operating budget was spent on "selecting recipients of [NIH] extramural grants."[9] That made NIH peer review an extremely expensive method of grant evaluation.

However, the committee concluded that the NIH's method "was an exceedingly good one" and said that as the NIH expanded, "peer judgments should be preserved at all costs."[10] The Wooldridge Committee report also pointed out that the study section opinions tended to override the opinions of NIH employees, writing that "the traditional reviewing procedure of the NIH was explicitly designed to place a large degree of responsibility on outside advisers drawn from the scientific community" and that NIH employees were "under rather strong restraint from participating in the decision-making process or even in negotiating with principal investigators over possible modifications of applications before submission to study sections."[11] That reliance on scientists' opinions was a point of pride for the NIH—but would soon become a point of contention with legislative stakeholders.

EARLY CONTROVERSY AT THE NIH: THE FOUNTAIN HEARINGS

In addition to the frequent internal reviews by the NIH itself, the NIH's method of awarding grants was the subject of a congressional investigation in the early 1960s after the Democratic congressman Lawrence H. Fountain of North Carolina raised concerns that NIH grants were being mismanaged. Fountain had become alarmed after examining some of the

NIH's appropriations requests and learning that scientists could adjust their grant budgets without needing NIH approval. He pushed for a closer examination of the NIH's finances, claiming that significant funds were being redirected to non-research expenses. Fountain's investigation brought him into conflict with the NIH director James Shannon, a confident, polished physiologist who first made his name as a researcher at the NYU School of Medicine. Initially Shannon seemed to agree that some grants might benefit from closer supervision and agreed to consider new supervisory rules. However, the NIH moved slowly to implement any changes. In March 1962, a frustrated Fountain assembled a subcommittee to evaluate the NIH's progress and called Shannon to testify.[12]

At the hearing, Shannon assured the committee that grant recipients "are selected on the basis of a rigorous screening by their scientific peers" and that "all subsequent administrative actions having to do with adjustments of budgets, and so forth, are essentially trivial in relation to this basic selection process."[13] He insisted that scientists themselves were the best judges of how grant money should be spent, and that the inner workings of the scientific community naturally safeguarded against fraud or excessive requests. In Shannon's view, the NIH's grant review process chose the best scientific applicants, and that in itself guaranteed that funds would be spent in a responsible manner.

However, an investigation by Fountain's committee had revealed that at least one NIH grantee had, indeed, mismanaged their funds. The NIH had made several grants to a private company called Public Service Research, Incorporated. In their required accounting documents, Public Service Research had included several expenses that were not strictly research-related, such as money to recruit personnel and funds for moving and renovating the company's offices. NIH policy did not permit those kinds of expenses to be charged to research grants, but the breach had gone apparently unnoticed and had not prevented Public Service Research from being awarded further grants. Although no one accused the company of intentional fraud, even Shannon was forced to admit that Public Service Research had misspent their awards and that the NIH should have noticed the problems before Fountain's audit brought them to light.[14]

The Committee questioned Shannon extensively about the NIH's selection process, asking him to walk them through exactly how the NIH chose which proposals to fund. At the end of that questioning, they seem to have concluded that the NIH's method of choosing grants via panel review was unproblematic. Instead, Fountain and his supporters blamed the NIH's accounting system for the problems. Fountain concluded that NIH grant spending should be much more closely supervised and that grantees should have to obtain approval before changing their budgets, and he rallied his fellow fiscal conservatives to pass legislation requiring NIH grantees to account for their spending more precisely. The result was new and much stricter accounting practices for NIH grant recipients as well as significant reorganization at the NIH. Post-award accounting was no longer the responsibility of the Division of Research Grants; instead, once grants were awarded, all spending would be overseen by a new and separate grants management division.[15]

It is significant that although many at the hearings agreed that Public Service Research probably should not have received multiple grants, no one in the early 1960s suggested that the NIH ought to rethink its grant reviewing method. They only argued that grant *accounting* should be more strictly monitored. Review by a select number of NIH-appointed experts was still considered an unobjectionable method of choosing which grants to fund. That, however, would soon change.

OMB VERSUS PEER REVIEW

NIH employees were understandably unsettled when, in spring 1973, they received a report from the OMB castigating the NIH's grant review methods. This was not a report that could be easily ignored. The OMB was a new federal office formed in 1970 as part of a reorganization of the Bureau of the Budget, and it reported directly to the president. The OMB was tasked with considering how agencies could best support the president's executive agenda—and it seemed that the Nixon administration felt the NIH was not doing all it could on that front.

The context for OMB's examination of NIH review procedures seems to have been President Nixon's high-profile "War on Cancer," a signature plank of his domestic policy that sought major advances in cancer prevention and treatment. The National Cancer Act (NCA) of 1971 tasked the director of the National Cancer Institute (NCI)—an NIH member institute—with establishing a national program of cancer research.[16] Between 1971 and 1973, the NCI's annual budged quadrupled.[17] The flow of funds into NCI had a quick, significant impact on the NIH's overall appropriations. The NIH's budget in 1970 was $1.06 billion; by 1973, it was $1.76 billion, a 70 percent increase.[18] Up until that point, the NIH had been given a great deal of latitude to choose which projects to fund and how to allocate its budget, which NIH leaders felt was appropriate. But the Nixon administration, concerned with the success of a signature policy initiative, felt that the legislative branch was entitled to a larger voice in how those new funds were being spent.

This clash between the administration's desire for control and the NIH's desire for continued autonomy led to conflicts with NIH leaders and the scientific community. Despite the financial support resulting from the NCA, relations between the Nixon administration and the NIH were far from easy. The NCI's budget was increasing, but the other divisions of the NIH were not seeing the same generosity—in fact, during this period many other member institutes experienced budget *cuts*. The NIH's postdoctoral training programs were the target of particularly dramatic cutbacks.[19] When Dr. Robert Marston, the NIH's director, objected to the administration's demand that he cut 10 percent from the budget for basic research grants, President Nixon fired him, leveraging a provision of the NCA that made the NIH directorship a presidential appointment.[20] That same month, President Nixon's science advisor, the engineer Edward David Jr., resigned, complaining that the Nixon administration was not taking his advice seriously.[21] Nixon did not seek a replacement; in fact, he dissolved the President's Science Advisory Committee entirely, declaring that he intended to rely on the NSF for scientific advice.[22]

Members of the scientific community took both Marston's firing and David's complaints as signs that science was out of favor with the Nixon administration. "The intentions of the White House planners for science policy are difficult to fathom," the *Science* newswriter Nicholas Wade wrote in an article on David's resignation. "The dramatic hirings, firings, and reorganizations of the last few weeks bear signs of ad hoc thinking."[23] Wade also noted that the OMB had been increasingly managing and overseeing federal research dollars. Although the Nixon administration's plans for science policy were still opaque, one thing seemed clear to Wade: The Nixon administration sought greater control over decision-making at the NIH, and they planned to use the OMB as one of the ways to get it.

The OMB's 1973 report, "The National Institutes of Health and National Institute of Mental Health Peer Review System," bore out that fear. The report began with an evaluation of what the NIH's founding legislation said about how the NIH should conduct grant review, concluding that "the only legal requirement related to the peer review system as it is presently established is that research grants must be approved by the appropriate national advisory councils."[24] With that legal stipulation established, the OMB report gave an account of how the NIH awarded its grants and then offered opinions on the problems inherent in the NIH system. The OMB listed eight perceived flaws in the way the NIH awarded funds. The major objection, the focus of two of the eight listed problems, was that the NIH's system placed too much power in the hands of the scientist-reviewers, both the ones in the study sections and on the advisory councils. It should be NIH administrators, the OMB argued, who held the power to judge and approve grants.

Other complaints indicated that the OMB report authors were uneasy with the entire concept of having scientists review grant proposals. The report argued that conflicts of interest were "unavoidable" in a system where a scientist could both serve as an NIH grant reviewer and apply for an NIH grant. The OMB report also argued that too much NIH grant money tended to concentrate in a small number of elite institutions. Finally, the OMB report characterized the criterion of "scientific merit" as insufficiently objective and criticized the two-tiered ranking system of "approved" and "disapproved" grants. Under this system, a grant could be "approved" by the initial review group but ultimately not funded; the OMB saw these "approved but unfunded" grants as a transparent attempt to pressure Congress for more NIH funding. They argued that NIH review-ers constituted a "clientele group" that had "a personal interest" in making sure the NIH budget increased every year.[25]

From there, the OMB report presented a list of changes that the unnamed authors felt could alleviate those problems. The possibility that the OMB report authors seemed to find most desirable was the first one they presented: Abolish the NIH's peer review system entirely, including not just the initial review groups but the advisory councils as well, and rely entirely on NIH administrators to decide which proposals would be funded. "This would be the ultimate in return to administrative control," the report declared, although its authors did acknowledge that there would be the significant obstacle of legislative approval since the NIH was legally required to employ the advisory councils.[26] If scientific expertise was necessary to evaluate proposals, the OMB said, the NIH could simply hire full-time proposal reviewers with PhDs.

The OMB's other suggestions were an eclectic mix. Alternative 2 was to prohibit reviewers from receiving NIH funds for a period after serving on initial review groups or advisory councils. Alternative 3 suggested that the NIH retain its initial review groups but "seek legislation" to eliminate the advisory council's role. Another suggestion was for the NIH to employ numerical rankings of scientific merit instead of "approve" or "disapprove" ratings. Finally, one of the suggestions supported a vision of NIH grant review that eliminated in-person panel discussions but retained the NIH's ability to consult experts on an "ad hoc" basis. The OMB further suggested that the NIH could conduct all reviewing by mail and select referees randomly rather than relying on established review groups.

The OMB's concerns and proposed solutions reflected a strong desire to keep decision-making in the hands of federal employees rather than outside scientific experts. From the point of view of the OMB, the NIH was outsourcing its responsibilities to scientists—people who were not accountable to either the NIH or the legislative branch in the same way that NIH employees would be. They saw the NIH's study section members as self-interested representatives of the scientific community, outsiders who would award money to themselves or to their friends if given the chance and who were actively trying to use their positions to pressure the government for more NIH funds. Allowing external reviewers to hold so much influence struck the OMB as an unacceptable abdication of duty on the part of NIH employees. And yet, there was also a sense in the OMB report that the NIH's expert advice was not external *enough*, that the recurring use of the same expert voices lent itself to abuse in a way that a system more like a journal's or the NSF's would not. If expert voices were going to advise the NIH, the OMB was more comfortable with having those expert voices kept out of Washington, DC, and consulted on an ad hoc basis about individual proposals rather than awarded an official appointment and given a pile of proposals to weigh in on.

More broadly, the OMB's distrust of scientists' advice reflected a shift in the status of science in the US. As we saw in chapter 3, the immediate post–World War II period was a high point for the cultural status of science and scientists. Presidents Harry Truman and Dwight Eisenhower brought scientists into the inner circles that developed policies on nuclear weapons; President Lyndon Johnson's Great Society programs hoped to leverage the social sciences to solve problems like poverty. Furthermore, despite the strong interest in government funding for science and in scientific advice for the government, the early Cold War was a period of deep skepticism about politicians having a say over scientific research. Stakeholders both within and outside US science argued that government control over science was a communist policy and often cited the infamous case of the anti-genetics agronomist Trofim Lysenko to argue that government interference in science would lead to the promotion of politically appealing but unscientific work.[27] US diplomats sought to win Cold War "hearts and minds" abroad by arguing that democracy, not communism, would yield the best science and technology, pointing to Soviet attacks on genetics as evidence.[28]

By the mid-1960s, however, Cold War tensions had begun to lessen (albeit temporarily), Lysenkoism was not quite as fearsome a specter, and more policymakers and members of the public were starting to question the technocratic assumptions on which systems like NIH peer review had been built. In 1966, it was revealed that the Congress for Cultural Freedom, an international cultural organization that had sponsored several large-scale conferences on science and intellectual freedom, was a CIA-funded project aimed at discrediting communist ideas. That revelation challenged the belief that the US was an environment where science was free from political interference.[29]

Furthermore, federally funded research—especially military research—was starting to become more controversial on US college campuses. Antiwar activists protested departments and research institutes that held large military contracts and, in some cases, persuaded universities to cut formal ties with research institutes that specialized in military work. Public hearings in 1965 prompted the cancellation of Project Camelot, a secret initiative that sought to use social scientific research to help US government operatives persuade Latin Americans to take up arms against communism.[30] In 1969, the student group "The Experiment" published an account of all of the military funding on Stanford University's campus, leading to a faculty vote banning classified research.[31]

An increasing number of scientists expressed concerns over the power of the "military-academic-industrial" complex in more dramatic ways. In 1969, MIT faculty members held a one-day strike intended to provoke, as they put it, "a public discussion of problems and dangers related to the present role of science and technology in the life of our nation."[32] A group called "Science for the People" disrupted several AAAS meetings in the early 1970s, shouting down speakers with military ties—including Glen Seaborg, president-elect of the AAAS and the discoverer of plutonium.[33] Combined with the ongoing clashes between biomedical researchers and the Nixon administration, such activities led to an environment in which US science and the US federal government were no longer seen as entirely harmonious partners, and in which the US executive branch increasingly perceived scientists as a group that was unlikely to support their goals.

IN DEFENSE OF PEER REVIEW

Both NIH leaders and scientists outside the agency came to the defense of NIH peer review, but the two groups did so using markedly different tactics. In their reply to the OMB, NIH leaders admitted that there were a few unusual aspects of the NIH's system. No other agency relied as heavily on panel review; most agencies used far more mail review. Furthermore, the NIH's system of entrusting review to a relatively small number of formally appointed experts who evaluated all proposals at once, rather than a wide-ranging list of unaffiliated referees who evaluated just a few proposals each, was unique.

NIH leaders defended those differences by arguing that the NIH's large size required a special system, writing that "the NIH, with the largest and most complex program of grant and contract support in its science field, has the most fully articulated and formalized procedures for deciding on merit."[34] NIH leaders also defended their use of external reviewers by arguing that hiring scientists to review proposals full time would be a fundamentally unworkable solution: "Even if the Agency were lucky enough to recruit such individuals, they would rapidly lose their competence if isolated from research and working investigators."[35]

However, the NIH response took strong exception to the charge that NIH administrators had little to do with decision-making. "The peer review system has the narrow function of assuring responsible management by providing high-quality advice as to the merit of individual projects," the director's response argued.[36] Reviewers did not allocate funds or even make final decisions on which proposals would receive money; they offered advice to the directors who would make those final decisions. The NIH's response to the OMB also claimed that despite some differences, the NIH's review procedures were not, in practice, all that different from ones at other federal agencies. "NIH practice follows that of larger agencies such as the National Science Foundation and the NASA Office of Life Sciences, in which the technical merit rankings are based on the opinions of outside experts," the response argued. "All agencies, *including the NIH*, rely on administrative decision-making for the allocation of funds. The steps in the process may differ, but agency program staff decide which projects shall in fact be funded."[37] These arguments were notably different from the 1965 Wooldridge report, which had characterized NIH peer review as heavily reliant on the opinions of scientists. The difference between that earlier report and the NIH's 1973 strategy suggests that NIH leaders were taking a cautious approach, not wanting to give credence to the OMB's description of their system lest they accidentally give credibility to their criticisms.

Scientists, however, did not shy away from describing the NIH's methods as special or different. Instead, they focused on making the argument that the NIH's "peer review" system was superior to other alternatives. The OMB report quickly made news among science policy analysts and science reporters. The *Science* news writer Barbara Culliton sounded an alarm about the OMB report in the magazine's columns, writing that the OMB sought to "dismantle NIH study sections."[38] US scientists quickly marshaled their support and began writing and calling government officials to defend the NIH's methods. "It has just come to my attention through Science Magazine that the peer review system for the awarding of research grants is threatened with dismantling," the psychiatrist

Sydney Pulver wrote to Caspar Weinberger, secretary of the Department of Health, Education and Welfare. "I am writing to plead for the utmost deliberation and consideration before such a drastic step is taken." Pulver wrote that his applications, including the rejected ones, had been evaluated with "objectivity, impartiality, and wisdom" by "top flight people in their field." He contrasted this with his negative experience with evaluations written by in-house staff members at other agencies: "Almost always the staff were not experts in the field I was working with, did not look for or value the opinion of experts and rejected or accepted applications on what seemed to me very flimsy grounds. Political factors played a major part in acceptance."[39]

Scientists also wrote to their representatives in Congress. In a letter to the New York senator James L. Buckley, Jozef Zwislocki, a laboratory director at Syracuse University, argued that "peer review is the only possible way of assuring that the most competent people doing the most promising research are funded. No other people are qualified to judge scientific work."[40] The biologist T. C. Hau at the University of Texas, the president-elect of the American Society for Cell Biology, told Representative Paul G. Rogers that getting rid of NIH peer review would result in a "waste [of] taxpayers' money on medical research of inferior quality." He argued that someone of his own stature would always be able to obtain grants "by playing politics with administrators," but that the NIH's review system made choices based on the quality of the proposal.[41] An official missive from the New York Academy of Sciences to the president argued that the NIH's peer review system "provides impartiality in selection of research projects of high scientific excellence. It allows experience and expertise of the scientific community to enter into the decision making process." The academy expressed the hope "that this system which is the envy of the world will be continued and reinforced," adding that peer review would surely bolster the president's national cancer plan.[42]

The NIH and the research community thus took very different approaches to defending the NIH's system. The NIH admitted that their system was a bit unusual but assured readers of their report that NIH administrators still played a large role in funding decisions. The scientific community, however, focused on a defense of the study sections—the part of the process that was designed to bring in scientific expertise and create a role in the process specifically for scientists. In doing so, they pushed back against the OMB's claim that scientists were self-interested and working to secure funds for themselves and their friends. Instead, they argued that scientists were uniquely impartial figures who cared only about the quality of a proposal, who could not be charmed through "playing politics." This language strongly echoed Robert Merton's "disinterestedness" norm of science, which said that for science to work properly, scientists needed to care for the truth of a finding over their own professional advancement. Scientists also strenuously argued that active researchers were the only people qualified to advise the government on whether a proposal had scientific merit, echoing language from science studies that cast review by fellow experts as the key to scientific reliability. Remove those experts, scientists argued, and the US government would find itself wasting money on inferior research.

THE RISKS OF NOT CONSULTING EXPERTS: THE 1973 "QUALITY OF HEALTH CARE" HEARINGS

The Nixon administration was quick to protest that they had no intention of dismantling NIH peer review. "If we were trying to eliminate peer review, I wouldn't be here," Charles Edwards, the assistant secretary for health in the Department of Health, Education, and Welfare told *Science* in an interview.[43] Indeed, Edwards even told *Science* that he was looking into the possibility of expanding NIH-style peer review to the FDA. But despite such attempts at reassurance, things between the Nixon administration and the biomedical community quickly became even more fraught. As part of the War on Cancer, the NCI was expanded and granted semi-autonomy from the larger NIH. Many within the biomedical community took that move as another attack on both the NIH itself and on the NIH's peer review system. In January 1974 John Sherman, an NIH deputy director who had been with the agency for over twenty years, resigned from the NIH and gave several interviews stating that he had done so because the Nixon administration's policies had led to declining morale and decreased confidence in the NIH's effectiveness.[44] Sherman described himself as "discouraged and disillusioned" and said that conflicts between the scientifically minded NIH and the parsimonious OMB were a major reason why.[45]

Sherman's complaints caught the attention of the editorial staff of the *Washington Post.* On January 19, the newspaper ran a short editorial on "Trouble at NIH," urging the Nixon administration and Edwards specifically to provide the NIH with "a different atmosphere and tangible support," including greater autonomy.[46] Edwards responded by accusing NIH leadership of making the NCI's removal inevitable: "If, in fact, the NIH leadership had been more perceptive and responsive we might not have witnessed the removal of the cancer research effort from the administrative control of the NIH." The NIH's future, he wrote, would be shaped by "many scientists at NIH and elsewhere who do not share the sentiments of Dr. Sherman" and who recognized "the need for constructive change."[47] Meanwhile, the clashes between OMB and NIH continued, in large part due to a perceived refusal on the part of the OMB to consider the advice of scientists when evaluating the NIH's spending. In an article for *Science*'s "News and Comment" section, Culliton wrote that "on the whole OMB and NIH are not on very good terms," observing that the budget manager assigned to the NIH was a recent college graduate with no scientific or managerial training.[48]

At the same time that the NIH and the biomedical community were wrestling with unwanted pressure from the OMB, however, a far more public and high-profile episode was raising serious concerns about biomedical research in the United States. Over a six-month period in 1973, the Massachusetts senator Edward Kennedy chaired a series of hearings on human subjects research, designed to probe into some recent and alarming medical experimentation. The hearings, which took place from February through July, covered a variety of studies on topics from electrical stimulation of the brain to human genetics research to the use of the long-term contraceptive Depo Provera.[49]

The hearings are most remembered, however, for shining a spotlight on the horrifying Tuskegee syphilis study, in which the US Public Health Service collaborated with the Tuskegee Institute in Macon, Alabama, to record the progress of untreated syphilis, a communicable and potentially fatal sexually transmitted disease. The subjects of the study were six hundred Black men, some who had the disease and some who did not.

The study began in 1932, before there was a known cure for the disease; patients were told that they were being treated for an unspecified blood ailment. By 1943, however, there was a well-known treatment: penicillin. The doctors running the Tuskegee study made the deliberate decision not to give penicillin to their test subjects in order to continue observing the progression of the untreated disease. Participants who knew they had syphilis were told that they were undergoing treatment when, in fact, they were merely being monitored for signs of the late-stage disease. The study continued until November 1972, marking four full decades of giving its sick subjects neither a clear diagnosis (or cautions on how to avoid spreading the disease to sexual partners) nor a widely available treatment to cure it.[50]

The scientists and policymakers called to testify at the hearings condemned the Tuskeegee study, and several of them argued that the reason it had moved forward was that it had never been subjected to review by qualified experts. At the hearings, witnesses argued that peer review could be a safeguard against future unethical studies. For example, Henry Simmons, the deputy assistant secretary for Health and Scientific Affairs at the Department of Health, Education, and Welfare (HEW), argued that one of the reasons the Tuskeegee study had been approved was that in the 1930s the department did not have "adequate peer review."[51] The situation for human subjects in biomedical research had improved "since the NIH mandate of peer review . . . for all the research it funded," the sociologist Bernard Barber argued in his testimony, although Barber added that there was still significant need for other institutions to enact regulations over human subjects research.[52]

Scientists and physicians at other agencies further reinforced the idea that peer review was a possible solution to unethical research when they testified that they were looking to expand their own peer review mechanisms. Charles C. Edwards, a physician serving as the Commissioner of Food and Drugs for the Department of HEW, said that the FDA's effectiveness in the future would depend on "adequate peer review" to provide oversight on which drugs were approved and how they might be prescribed.[53] Kennedy himself held up the NIH's system as a model for other federal agencies to emulate. This was no accident; Kennedy was known as a friend of the NIH and had taken steps to publicize the OMB's attacks on NIH peer review with the goal of marshalling support for the NIH's system.[54] The 1973 hearings thus became not just an important moment to consider how to prevent unethical and discriminatory biomedical research, but an occasion where peer review was held up as a way to protect science from government interference and to protect the public from government incompetence or ethical failure.

CONCLUSION

Despite the alarm it occasioned for the NIH and the scientific community more broadly, the OMB report on NIH peer review ultimately did not lead to significant changes at the NIH. One major reason was that the Nixon administration was distracted by a slightly more pressing political matter. January 1973, the month that the OMB released its report, was also the month that trials began for two former Nixon associates accused of breaking into and attempting to wiretap the headquarters of the Democratic National Committee at the Watergate Hotel.[55] Nixon had steadfastly maintained that he had no personal knowledge of the break-in, but a slow trickle of revelations soon made it clear that there had been a concerted (and criminal) effort to cover up the president's involvement. The embattled Nixon resigned the presidency on August 8, 1974—the first US president to do so.

The OMB report's significance lies less in its impact on federal spending than with its impact on the development of language that defended and justified the use of a "peer review" system. Faced with an administration that many biomedical researchers saw as hostile to scientific advice, scientists and NIH employees began arguing that the NIH's "peer review" was superior to other methods of reviewing. Government employees did not, *could* not, have the same qualifications to review proposals as working scientists—review had to be done by the "peers" of the person proposing the research. The OMB had argued that scientists' opinions could not be trusted because those scientists were self-interested; in response, defenders of the NIH's review system argued that scientists were in fact uniquely objective observers whose sole concern was scientific merit, echoing Merton's "disinterestedness" norm of science. In the "Quality of Health Care" hearings, scientists and legislators argued that lack of expert review had led to research that was not just unscientific, but unethical and harmful. The OMB had argued that administrative review would render the NIH more accountable to the legislative branch; scientists responded by arguing that it was the *government* that needed to be kept accountable, and that peer review by scientists was the mechanism that would protect both US science and the public at large.

In these early 1970s conflicts, we begin to see scientists argue that peer review was not just one option among many for evaluating scientific projects; it was the *only* method that could lead to high-quality and ethical science. By the mid-1970s, US scientists had begun to see "peer review" as a process they could cite and defend in order to prevent the US government from interfering too deeply in the workings of science or cutting scientists out of decision-making at funding bodies like the NIH. We will see that sort of language on a larger stage in our next chapter, which focuses on a controversy over peer review at the National Science Foundation—a much higher-profile episode than the OMB-NIH clash, and one that significantly reshaped wider perceptions about peer review's role in science.

6

"THE BEST METHOD"

THE 1975 NSF PEER REVIEW HEARINGS

The NIH's allies had successfully defended the agency's unusual "peer review" system, but all was not well in US science when the controversy drew to a close. In chapter 3, we saw the US government of the 1950s funding scientific research at an unprecedented level, dwarfing the spending from private funding bodies and altering the scale of job opportunities, graduate training, and journal publishing. By the early 1970s, the situation had dramatically changed. The technocratic optimism of the 1950s had faded and, as we saw in chapter 5, scientists were increasingly the subject of cultural and political scrutiny rather than instinctive respect. The US economy entered a brutal period of stagflation—high unemployment combined with high inflation and stagnant growth—and federal spending on science began to decline for the first time since World War II. That decline in spending jeopardized job opportunities not just in the federal government but at companies and universities that had become dependent on generous federal grants. Scientists had good reason to wonder about their social position, not to mention their job security.

Just two years after the OMB-NIH dustup, a series of controversies surrounding grants at the National Science Foundation provided further evidence that scientists should be worried. Lawmakers from both major parties accused the NSF of wasting taxpayer money on frivolous, nonsensical projects with no value to the American people. The NSF's method of choosing which grants to fund relied heavily on advice from scientists, and suddenly, congressmen and senators were demanding copies of grant referee reports and suggesting that the NSF needed to change its approach. It bore remarkable similarity to the NIH's troubles with the OMB, but this time, the stakes were higher. The OMB-NIH clash had drawn only mild interest from the public, but thanks to the press savvy of the NSF's legislative critics, NSF grants were making headlines in national newspapers and syndicated columns.

From this unpromising situation, however, US scientists managed to score a victory. They not only preserved their role in giving advice to the NSF but they also won converts to the idea that review by scientists was a necessary part of grant-making that had to be preserved in order to ensure the quality and integrity of federal spending on science. Scientists and NSF leaders argued that the NSF had to solicit advice from scientists because disinterested "peer review" was the key to properly working science. The argument worked; in fact, it worked so well that other federal agencies began leaning more heavily on peer review, worried that they might be criticized for taking other paths. Even some of the NSF's harshest critics agreed that review by a scientist's peers was the best, most legitimate way of deciding which grants to fund. By the late 1970s, the idea that "peer review" was central to science was widely accepted by stakeholders outside the scientific community. Peer review, in other words, had gone mainstream.

THE PURSE STRINGS TIGHTEN

As we saw in chapter 3, federal spending on science seems to have been largely uncontroversial in the 1950s and early 1960s, largely because of the perceived need to keep up with the Soviets. Several new developments in the mid- to late 1960s, however, called that federal largesse into question. In 1966 the Department of Defense (DoD) issued the Project Hindsight report, the result of an investigation into the outcomes of DoD funding for research in the physical sciences. The Hindsight report concluded that while "the DOD investment in science and technology has had a large payoff," the payoffs were "delayed in time by 5 to 10 years."[1] Furthermore, outcomes were far better "when funded and managed by the

Defense Department or defense industry" rather than "funded and managed by the non-defense sector of government or industry."[2] The most significant conclusion, however, was that outcomes were best "when the effort was oriented"—that is, when money was spent on applied research rather than on what the report called "undirected research."[3] Although the Hindsight report was specifically about physical sciences research and weapons systems, the report led many policymakers to question military and government expenditures on basic research more generally, on the grounds that "undirected" work seemed less likely to yield direct benefits for the American public.[4] The US federal government's focus shifted from creating a "standing army" of new scientists to sponsoring projects that would create useful new technologies or directly support political goals.

The post-Sputnik fear of Soviet science had also lost much of its power. In 1969 the Cold War entered its period of détente, and tensions between the US and the USSR temporarily abated. This blunted one of the most persuasive arguments in favor of increased science funding—namely, that the US had fallen behind the Soviets and needed to catch up. Many Americans took the 1969 moon landing as evidence that the Americans had regained a now-insurmountable lead over the USSR in terms of science and technology.[5] The US cultural view of scientists was also changing. In the decades immediately after World War II, scientists were riding a new wave of prestige and trust, consulted by lawmakers and bureaucrats across government for their expertise. But by the late 1960s, as we saw in chapter 5, skepticism of science was on the rise on both sides of the political spectrum. Left-wing activists were angry about the scale of scientific collaboration with the US federal government during the Vietnam War; conservatives were angry that left-wing scientists had voted classified research off campuses and participated in anti-war activities.

The combination of changing government priorities and bipartisan disappointment in science created the perfect circumstances for cutbacks to science funding. In the late 1960s, for the first time since the end of World War II, federal funding for science began to experience a decline (adjusting for inflation).[6] The consequences of the shrinking funding for American science were quick and dramatic. Students who had entered PhD programs in the early 1960s felt confident that they would have robust job opportunities after finishing their doctorates. In 1962, for example, the Placement Service at the American Institute of Physics (AIP) listed 514 jobs for physicists but only 449 PhDs looking for work. Just six years later, however, the 1968 Placement Service records listed 989 applicants and 253 jobs.[7] A 1969 *Physics Today* article reported that 40 percent of new graduates with physics masters' degrees and 30 percent of new physics PhDs had received zero job offers.[8] By 1971, the numbers were outright catastrophic: The AIP Placement Service registered 1,053 applicants looking for work, but advertised just 53 jobs.[9]

"MAKE A MONKEY OUT OF THE AMERICAN TAXPAYER"

The resignation of Richard Nixon in 1974 provided scientists with a brief moment of optimism. President Gerald Ford seemed determined to lower the temperature in Washington, and scientists were among those expressing relief at the transition to a president with, as an article in *Science* put it, "a reputation for openness and fair dealing."[10] *Science* also praised Ford for convening a scientific advisory committee composed of scientists from his district during his time as a congressman. The physicist Vernon Ehlers, who had served on that advisory committee, called the new president "very astute" and said that he "displayed a good grasp of the subtleties of scientific issues."[11] Ford restored both the office of scientific advisor and the President's Science Advisory Committee, which Nixon had eliminated following the resignation of Edward David Jr.

If the scientific community was hoping that the Ford administration would show more largesse toward scientific research, however, they were quickly disappointed. "Biomedical Research Budget Drops Again," *Science* reported in February 1975, noting that Ford's budget could mean a $285 million budget cut for the NIH in the span of just a year.[12] However, there was little sense that the NIH was being specially targeted. When Ford took office, the US economy was facing major challenges. The combination of an oil crisis and stagflation led to widespread unemployment, a decrease in real wages, and budget restrictions at a wide range of organizations. *Science*, for example, adopted strict page limits in response to declining AAAS membership revenues and increasing paper prices.[13] The federal government was under particularly intense pressure to cut its budget in the face of a decreased tax base and widespread economic anxiety.

By 1975, two Republican congressmen and a Democratic senator had set their sights on the NSF as a potential target for budget cuts. Congressman John Conlan (a Republican from Arizona), Congressman Robert Bauman (a Republican from Maryland), and Senator William Proxmire (a Democrat from Wisconsin) each had distinct reasons for questioning the NSF's budget and grant-making procedures. Conlan was a first-wave member of the Christian Right who was gearing up to run for the Senate, billing himself as a social conservative who would bring Christian values to his government service.[14] Two NSF-funded school programs had become controversial among Arizona's social conservatives: Man, a Course of Study (MACOS) and the Individualized Science Instructional System (ISIS). MACOS was a social sciences curriculum that covered the social habits of several different species of animal, as well as the Indigenous Netsilik people in northern Canada. Critics such as the Heritage Foundation, a conservative think tank, accused MACOS of promoting moral relativism and focusing on violence, incest, and cannibalism among animals.[15] ISIS, a program aimed at fourth graders, also drew scrutiny for allegedly including too much explicit discussion of human sexuality.

During congressional discussions about the NSF's 1976 appropriations request, Conlan stridently criticized MACOS and ISIS and questioned whether the NSF was spending its budget in the interests of the American public. Conlan soon found an ally in Bauman, a small-government advocate and a founder of the American Conservative Union who was then a rising star in the Republican party.[16] Bauman proposed an amendment to the 1976 NSF Appropriations request requiring the NSF to receive congressional approval for any grants it wanted to fund. The amendment was approved and added to the appropriations bill.

It was Proxmire, however, whose NSF criticisms made the biggest public splash. Proxmire had been a senator for nearly two decades. He won a special election in 1957 to replace the recently deceased Joseph McCarthy and cast his first vote in favor of the Civil Rights Act after the act's opponents made a brief and ultimately fruitless attempt to prevent him from being sworn in before the bill came to the floor. The senator quickly made his mark as a combative legislator with a penchant for cutting press releases and principled, stubborn stances. He was an early opponent of the Vietnam War, an early advocate of campaign finance reform, and he is still the record-holder for most consecutive roll call votes (his attendance streak at Senate roll call votes stood at 10,252 when he retired in 1988).[17] He was also the author of several lively books, including *Report from Wasteland: America's Military-Industrial Complex*; *Uncle Sam: The Last of the Bigtime Spenders*; and *You Can Do It! Senator Proxmire's Exercise, Diet, and Relaxation Plan*.

Today, Proxmire is best remembered as a fierce critic of federal spending—and for his skill at calling reporters' attention to projects he found wasteful. In March 1975, Proxmire's office issued a press release describing the senator's new "Golden Fleece Award." This "award" was actually a badge of dishonor that Proxmire planned to bestow upon the month's worst use of federal taxpayer dollars. The March recipient? An NSF-funded sociological study about interpersonal attraction by Ellen Berscheid at the University of Minnesota and Elaine Walster at the University of Wisconsin. "I believe that 200 million other Americans want to leave some things in life a mystery, and right on top of the things we don't want to know is why a man falls in love with a woman and vice versa," Proxmire proclaimed. "So National Science Foundation—get out of the love racket."[18] The second Fleece in April 1975 also went to an NSF-funded study, the psychologist Ronald Hutchinson's work on why humans, rats, and monkeys clench their jaws in moments of stress. "The funding of this nonsense makes me almost angry enough to scream and kick or even clench my jaw," declared the month's Golden Fleece press release. "The good doctor has made a fortune from his monkeys and in the process made a monkey out of the American taxpayer."[19]

Proxmire's Golden Fleeces won widespread media coverage, and letters complaining about "questionable" grants began flooding the NSF offices. The critical letters were often forwarded to the NSF by the office of a senator or a congressman and, as such, required a reply. The NSF quickly developed a standard response to these inquiries, which was to assure constituents that the peer review system had been thoughtfully developed and that the NSF only funded proposals of the highest quality.[20] Some of the letters to the NSF complained about grants that other government agencies had funded, a misunderstanding that highlighted the intensity of the scrutiny on the NSF in particular.[21] The researchers Proxmire had targeted were also singled out for criticism and angry mail.[22] The Walters-Berscheid grant was not renewed, and social scientists felt Proxmire was to blame.[23] In a lawsuit filed in 1976, Hutchinson said he had suffered mental anguish and public humiliation, and accused Proxmire's staff of deliberately interfering with his ability to secure other government grants and contracts.[24] Proxmire, apparently undeterred by Hutchinson's lawsuit, would make the NSF a favorite target of his Golden Fleece Awards, insisting that the organization funded frivolous and wasteful projects that were of no use to the American people.[25] And the NSF could not dismiss the Wisconsin senator's criticisms as mere showboating; throughout the 1970s, Proxmire sat on the Senate's Finance Committee, giving him oversight over NSF appropriations.[26]

The school curriculum controversy—in particular, the criticisms of MACOS—also received a great deal of press coverage. The syndicated columnist James Kilpatrick was a particularly visible media detractor. In one March 1975 column, Kilpatrick accused the NSF of "promoting [a] particular ideology" with their curriculum grants and said that "MACOS is a subtle and skillful exercise in behavioral manipulation." The NSF's legislative critics, Kilpatrick declared, were "asking hard questions that should have been asked of the NSF long ago."[27] In later columns, Kilpatrick declared that "Conlan is on the right side of the fundamental issue," encouraged parents to "put pressure on their own congressmen,"[28] and argued that federal funding for textbooks was "an ominous echo of the Soviet Union."[29]

But the controversy was not only being carried out in the media. Dozens of letters were exchanged between the NSF, Proxmire, Conlan, and Bauman as the three legislators sought further information on specific NSF grants. By March 1975, the controversy had grown so intense that the NSF director, H. Guyford Stever, announced that no further funds would be approved "either for MACOS or any other precollegiate science course development . . . until we have conducted a thorough review of the NSF effort in these areas and reported to the National Science Board and Congress with recommendations."[30] That decision did not mollify Conlan, who was determined to learn how the NSF had decided to fund ISIS and was frustrated when the NSF refused to give him copies of the grant's peer review reports. "I would again remind you that I am a Member of Congress on a Committee charged with the oversight of the

National Science Foundation," he wrote in a May letter to Stever. "Consequently, I do again demand that you make available the peer reviewer comments originally demanded by me—in their original and complete form, not paraphrased."[31]

NSF leaders, however, argued that peer review depended on the anonymity of the referees and said that the NSF was not obligated to disclose the reports to anyone, including members of Congress. In a letter to Conlan, Stever said that referee reports were filed under an "implied promise of confidentiality" that was "well understood and accepted by thousands of reviewers," and that releasing the text of the reports or the names of the reviewers would constitute a breach of trust.[32] He further argued that any change to the NSF's policy of keeping referees anonymous would be at odds with the concept of peer review itself:

> Maintaining the confidentiality of comments solicited by the Foundation on grant proposals and the identity of the reviewer making these comments is a fundamental principle on which the Peer Review System of the National Science Foundation and also of other agencies is based. . . . If this confidentiality principle were to be changed, the entire process whereby the wisdom and knowledge of outstanding scientists and educators throughout the country is drawn upon in evaluating the worth of proposals would thereby be changed.[33]

Confidentiality would continue to be a bone of contention between the NSF and its critics as the conflict progressed.

The careful phrase "implied promise of confidentiality" suggests that instructions given to NSF reviewers did not always explicitly promise that the reports would be kept confidential; however, Stever's response to Conlan was consistent with internal NSF policy. By 1975, according to internal NSF statistics, 44 percent of proposals were evaluated by ad hoc mail review, 28 percent received panel review, and 28 percent received reports from a mix of the two systems.[34] Specific instructions to reviewers have not been found in the NSF archives, but NSF correspondence with both grant applicants and Congressmen routinely stated that it was the NSF's practice not to disclose the full text of the referee reports or the names of the reviewers. "[Referees] are private citizens who serve voluntarily without compensation, often devoting several hours of their own time to review the proposal," the NSF general counsel Charles F. Brown told one disgruntled applicant. "If the name of a reviewer were made public, the reviewer might be subject to harassment or badgering by the principal investigator."[35]

THE NSF HEARINGS: SCIENTIFIC AUTONOMY OR PUBLIC ACCOUNTABILITY?

The public debate and behind-the-scenes arguments over the NSF's funding decisions led to the National Science Foundation Peer Review Special Oversight Hearings, held before the Subcommittee on Science, Research and Technology in July 1975. Bauman, Conlan, and Proxmire were not members of the committee but were invited to testify at the hearings, as were Stever and the NSF deputy director Richard Atkinson. Over a dozen scientists, sociologists, and other academics were also asked to testify about their opinions on peer review. The hearings generated nearly 1,200 pages of transcripts and supplemental documents. These transcripts contain not only discussion of the NSF's particular review process but much discussion of the concept of peer review itself—its outcomes, its ideal form, and its role in scientific practice.

The discussion of the NSF's review process focused on two key points: first, what criteria should be used to choose which NSF proposals to fund, and second, how much weight should be given to the referees' opinions about a proposal. During questioning, both Bauman and Conlan acknowledged that their interest in the NSF stemmed from concerns that some of the funded proposals were not beneficial to the American public. Bauman argued that "we are dealing here with a finite quantity of money taken in the form of taxes from people against their will in a very difficult time in our economy and being spent by a Federal agency, spent in a manner now questioned," and that the NSF "must demonstrate that it is using tax money in a prudent manner, in such a way that the taxpayers can expect that there will be some payoff from NSF-funded research."[36] Bauman argued that NSF had strayed from its mission of funding the best basic research in the natural sciences. He provided the hearings with a list of recent grants that he deemed suspicious and took specific aim at the NSF's funding of the social sciences, questioning "whether they should be classified as sciences at all."[37] His list of questionable grants included many in economics, political science, and science communication, as well as a $130,700 grant to the Columbia sociology department to study sociology of science.[38]

Conlan's complaints were more focused. Much of his testimony centered around his objections to MACOS and ISIS and his frustration at the NSF's refusal to provide him with the full, verbatim referee reports. Conlan described the NSF's peer review system as a secretive process that allowed program directors to elicit any decision they wanted:

> Here is an amazing system, gentlemen, where individual program managers are given carte blanche authority to select peer reviewers . . . [It is] a completely arbitrary system that is closed and unaccountable to the scientific community and to the Congress. . . . It is common knowledge in the science community that NSF program managers can get whatever answer they want out of the peer review system.[39]

Significantly, Conlan's criticisms of the NSF did not lead him to criticize the concept of peer review in general. Instead, Conlan argued that the NSF's review system was designed to allow the directors, rather than the reviewers, to make decisions about the quality of a proposal. Because no one outside the NSF was allowed to see the reports, said Conlan, neither scientists nor congressmen could verify that NSF staff members were listening to the referees. Conlan claimed that NSF program managers occasionally went so far as to misrepresent their referees' comments. He recounted a conversation with one of the ISIS peer reviewers, Philip Morrison of MIT, and reported that Morrison had been very critical of the proposal and was surprised to learn it received funding.[40]

Conlan therefore argued that all peer review reports, for funded and unfunded grants, should be available to Congress and to grant applicants. In Conlan's view, this would stop the NSF from making any further questionable grants: "If peer review and NSF grants management are open, I think the area of faulty grantmanship or inadvisable or incompetent awards will take care of themselves."[41] Conlan also suggested that the reviewers' names ought to be available to applicants, arguing that this would encourage more objectivity in the refereeing process because reviewers would work harder to be fair if their names would be attached to their reports.[42]

Stever vigorously defended the work of NSF employees, arguing that their expertise and professionalism were crucial to the NSF's success.

However, he said, "it is the Foundation's internal practice to make sure that no member of its professional staff, no matter how competent, is permitted to make a decision in private on behalf of the Foundation." Stever attributed the Golden Fleece and curriculum controversies not to flaws in the NSF's review process but to a gap between public expectations and scientific realities. "Society is expecting more of science," he said. "Some of these expectations are reasonable, but some are unreasonable expectations of magical results on a faster timetable than ever will be realized."[43] Because the benefits of scientific research would not always be immediate or obvious, said Stever, the foundation believed that proposals should be judged based on the quality of the science. "The foundation never supports frivolous research," he insisted. "Scientific excellence is the primary criterion for NSF support."[44]

The only way to choose proposals based on scientific excellence, Stever and Atkinson argued, was through a peer review process—by placing proposals in the hands of those who were best qualified to know what scientific excellence looked like. The foundation consulted with "members of the scientific community who are expert in the specialty or discipline involved before making a recommendation to support or to decline a specific proposal," Stever explained. "The persons consulted are expected to be the scientific peers of those proposing to do the research; thus the growth of the term 'peer review.'"[45]

Stever repeatedly insisted that these peers could only give candid advice if their identities were kept confidential; otherwise, reviewers might fear professional reprisal if they did not recommend particular proposals, especially those by senior colleagues in their field. Referees could only be free to speak their minds, he said, if they knew they would remain anonymous. Stever's comments cast the referee as a person who was performing a selfless service to science with no hope of reward, but also as a potentially malleable figure who might succumb to external pressure if their identity became known.

This disagreement over anonymity reveals a major difference between the NSF's critics and its supporters. NSF leaders supported anonymous peer review because, they argued, it was the only way to ensure candid and accurate feedback about proposals. Legislative critics, on the other hand, felt that the NSF's decisions should be more transparent to legislators, taxpayers, and the scientists who applied for grants. In other words, the NSF's priority was to promote scientific candor; the critics' priority was to make the process accountable to Congress and to the public.

SCIENTISTS IN DEFENSE OF PEER REVIEW

The scientists called as witnesses were an eclectic group of voices from a range of institutions, fields, and career paths. (They did share one notable thing in common: With a single exception, the University of Virginia engineer Doris Kuhlmann-Wilsdorf, they were all men.) They did not speak with a single unified voice. Many of them expressed frustration with particular NSF peer review outcomes or suggested adjustments that they felt would improve the NSF's process. However, they agreed on one point: The NSF needed to fund the best science, and the only people who could determine the scientific merit of proposals were scientists themselves.

John F. Sherman, a medical researcher and former deputy director of the NIH who had been at the agency during the OMB controversy, called peer review the "best available method" for evaluating the scientific quality of grants. A peer review system that relied on expert advice from scientists, said Sherman, gave a federal agency "a necessary degree of protection against influences which, however well intentioned, would undoubtedly result in less effective use of public funds."[46] William P. Raney, the chief scientist at the Office of Naval Research, strongly argued for the role of scientific experts in determining which projects should receive government money. The mechanical engineer and Catholic University vice president Donald Marlowe described peer review as "a widely understood and respected system" with "deep historic roots." Although he admitted the possibility of human failures, Marlowe insisted that "it is difficult to imagine any other system for bringing the great variety of required expertise to bear on the critical evaluation of proposals." Peer review, he said simply, "is a practical system—it works!"[47]

The influence of the sociology of science over the discourse about peer review is also clear in the text of the NSF hearings—a slight irony, given how controversial the funding of social sciences was at other points in the hearings. Scientists called as witnesses cited the work of figures such as Kuhn and de Solla Price to back up their claims about how science should operate. Some of the strongest voices in support of the peer review system were social scientists themselves. Charles Kidd, the executive secretary of the Association of American Universities, had no formal natural sciences training; he held degrees in history, economics, and public administration. But he was unequivocal in his support for the peer review system. Testifying on behalf of "major national associations representing institutions of higher education," Kidd argued that for the past several decades, "the peer review system has survived searching examination. Every serious review by probing, independent groups has ratified the essential utility and equity of the system. . . . The system has stood the test of time and the test of critical examination."[48]

Another strong supporter of peer review was Raymond Bowers, a professor of physics at Cornell University who was also the director of Cornell's program on science, technology, and society, as well as a noted expert on science policy. Bowers knew DC politics well; he was a former staff member at DC's Office of Science and Technology. That experience was reflected in his testimony. Unlike many of the scientist-witnesses, whose insistence on nuance led to long answers and perplexed follow-up questions from legislators, Bowers's responses to questions were clear, direct, and concise. He left no doubt as to what he thought of the peer review system: "I am convinced that peer review is an indispensable component of the decisionmaking process for allocating funds."[49] Bowers also argued forcefully for reviewer anonymity. "I am very skeptical about the potential quality of the reviews that would result if the reviewer knows his identity will be made public," he said. "We have all had experience with those letters of reference carried by a potential employee in an open envelope. Have any of us ever seen such an appraisal suggesting that the person is not competent?"[50]

The defense that scientists offered of peer review was both theoretical and practical. Supporters argued (somewhat ahistorically, as we've seen) that peer review had been a standard part of science for decades and that its role in science was well understood and widely accepted. But they also claimed that peer review had a central role in ensuring fiscal responsibility. Without evaluation by experts, they argued, federal money might be spent on truly worthless projects, ones that had no chance of success.

Furthermore, the identities of the evaluators had to remain unknown to applicants. If the names of the scientist-reviewers were known, fellow scientists might seek to influence the reviewers, leading to recommendations based on friendship or the status of the applicant rather than on scientific merit. Despite any individual cases of flawed decisions, peer review as practiced by the NSF was the only possible way of choosing between grant proposals.

"WE ASK TOO MUCH OF IT": THE LIMITS OF PEER REVIEW?

Not everyone at the 1975 hearings, however, was convinced that peer review would secure grants for the best science. The Pennsylvania State physicist Rustum Roy, known for both his materials science work and his writings on science policy, was one of the few witnesses to point out that peer review was (as we saw in chapter 5) not a universal practice at funding agencies. "We have many systems for science funding, besides peer review," Roy observed in his testimony, pointing to both other US federal agencies and contract-based funding systems in Japan and West Germany.[51] He suggested that the system of scientists sending proposals in whatever field they wished should be replaced with agency solicitations of proposals on specific questions, and that panels of preselected experts should always be used instead of ad hoc external peer reviewers.

Perhaps most evocatively, William D. Carey, the executive officer of the AAAS, argued that Congress and the NSF might be asking too much of peer review. Peer review, he told the committee, "is not a fail-safe procedure, nor one that should silence disagreements on the theory that there is something sacramental about it." Carey endorsed the older system of using NSF peer review reports as advisory documents, saying, "Peer review has its uses as a first round of proposal screening, but it does not absolve the Government program manager from full responsibility for the decision to fund or reject a proposal. . . . There are some things that we should not ask of peer review. We should not ask it to take Government agencies off the hook on the question of protecting the public purse."[52]

It was Proxmire, however, who expressed the most negative ideas about peer review. In his submitted testimony, he wrote, "I have received a number of letters pointing out that the peer review system serves to perpetuate the funding of established people, ideas and institutions." Proxmire seemed to hold out little hope that peer review could be anything other than "incestuous." He argued that NSF reviewers would favor proposals whose research had been published in peer-reviewed journals, peer-reviewed journals would look favorably on articles that had received NSF funding, and that "we come full circle, when we realize that many of the top researchers in a scientific field have been recipients of NSF grants, reviewers of NSF grants and finally editors of their technical journals."[53]

But interestingly, any limitations of peer review received little discussion during the testimony of NSF officials. The legislators on the subcommittee did not take this up as a line of questioning. Almost all of the actors at the 1975 hearings agreed that peer review was the appropriate mechanism for evaluating grants. Peer review, according to those at the hearings, was "a fair and effective system,"[54] "an indispensable component of the decisionmaking process for allocating funds,"[55] and "an integral feature of 'the scientific method.'"[56] Subcommittee chair James Symington seemed to sum up the hearings' attitude toward peer review when he said:

> Witnesses overwhelmingly agree that some form of peer review should continue to be used to assist in the allocation of funds for scientific research. To be sure, no witness—including the NSF—has claimed that the Foundation's peer review system and its decisionmaking ability are perfect. . . . Nevertheless, it appears that, as a base to work from, the peer review concept is seen as fundamentally sound.[57]

In 1973, government officials had expressed serious doubts over continuing to use peer review; by 1975, peer review was seen as a crucial part of federal grant-making in the United States.

OUTCOMES FOR THE NSF—AND FOR THE NIH

NSF leaders came to the hearings prepared to make changes that they hoped would mollify critics. Most significantly, at the hearings, Stever announced that as of January 1, 1976, NSF applicants would receive complete, verbatim copies of their referee reports with the reviewers' names redacted. Stever argued that keeping the referees anonymous was the only way to ensure their candor and incorruptibility, but he offered the verbatim reports as a compromise.

Stever and Atkinson also indicated that, in the future, the NSF would lean more heavily on peer review, and less heavily on the judgment of NSF staff, when deciding which grants to fund. A new audit office would ensure that funding decisions placed appropriate weight on positive and negative referee reports. Finally, the NSF commissioned a report on their peer review process from the RAND corporation to learn more about scientists' views on the efficacy and fairness of their process.[58] Despite such assurances, the NSF did not emerge from the episode entirely unscathed. In the wake of the controversy, the NSF's education programs were downsized, and funding for MACOS and ISIS was largely eliminated. However, Conlan and Bauman did not get the congressional oversight of NSF grants they had wanted, and the Bauman amendment was removed from the NSF's appropriations request.

Meanwhile, although the NIH was not directly involved in the 1975 controversy, NIH leadership—including new director Donald H. Fredrickson—watched the debate closely.[59] Sherman's testimony briefly put NIH peer review front and center at the hearings, and NIH leadership readied itself to respond to any criticism that might come its way. Perhaps hoping to head off attacks, Frederickson created a Grants Peer Review Study Team chaired by Dr. Ruth Kirschstein, director of the National Institute of General Medical Sciences, to analyze the NIH's methods.

The team's 1976 report to the director on "Grants Peer Review" made recommendations for new policies on conflict of interest, new methods of selecting council members, and the training of new study section members. The report overall lauded the NIH's system; their recommendations were designed to preserve reviewer privacy, limit workloads, and create options for the NIH to consult even more experts on an ad hoc basis without requiring an appointment to a study section.[60] The recommendations Frederickson eventually adopted included restating a commitment to "dual review," keeping advisory group meetings closed to the public, and confirming the director's right to create new review groups in order to manage reviewer workloads.[61] Like the NSF, the NIH also began providing more detailed comments about proposals to NIH applicants.[62] Starting with the decisions from the September–October meetings of the NIH Council, principal investigators received a summary statement containing feedback on their grants with their priority scores displayed.[63] The new rules and regulations released in 1978 reaffirmed that "peer review" had to be done by scientific experts:

"Peer review group" means a group of experts qualified by training and experience in particular scientific or technical fields to give expert advice, in accordance with the provisions of this part, on the scientific and technical merit of grant applications or contract projects in those fields.[64]

The new regulations also mandated that the NIH would use peer review groups to review not just grants for basic research but also contracts that required the achievement of specific scientific goals—a major change to the previous system for awarding contracts, which had permitted an eclectic range of review procedures.[65]

Notably, both the NSF and NIH's reforms placed *more* emphasis on the opinions of external referees. In 1973, OMB critics had argued that funding body employees were the most trustworthy people to review proposals because they would know their jobs were at stake. In 1975, the NSF's defenders argued that funding body employees had insufficient scientific training to judge applicants. The judgment of anonymous referees, on the other hand, was high-quality advice that could serve as a stand-in for the judgment of the scientific community as a whole. Peer review, and peer review alone, could be relied upon to produce answers about what was and was not a high-quality proposal. It was a vision of peer review that strongly echoed both Zuckerman and Merton's claim that journal refereeing was the system that ensured the reliability of science. It also invoked Merton's norm of disinterestedness, pushing the idea that for peer review to work, the reviewers had to be anonymous figures who did not stand to benefit or suffer in any way from their remarks. Ideas about refereeing that had first been applied to journals had officially been brought over to the realm of grant-making.

TOWARD THE 1980s

"The success of the peer-review system, and therefore much of contemporary science, depends on the help provided by the numerous scientists who serve on NIH public advisory groups," George Eaves, the executive secretary of the NIH's Molecular Biology Study Section, wrote in 1978. "These men and women are willing to give their valuable time because they recognize that the quality of science in this country depends on constant surveillance and guidance by the scientists themselves. They also recognize the probable consequences were they not afforded the opportunity for active participation in the policies and decisions related to the public support of research."[66]

Eaves's words nicely capture the general discourse about peer review in the US at the end of the 1970s. Review by scientific experts was no longer seen as an optional or occasional part of science; it was spoken of as a crucial part of both scientific knowledge-making and US federal science policy. The increased centrality of refereeing to scientific practice was also noticeable at scientific journals. As we saw in chapters 3 and 4, the use of refereeing at US scientific journals had been steadily expanding in the 1950s and 1960s, but by the late 1970s, refereeing for a published article was seen as mandatory. *Cell*'s founding editor Benjamin Lewin, for instance, recalled that when the journal began publishing in the mid-1970s, "there was never any question about it [using referees] . . . It was absolutely a given that you do not publish anything without peer review from somebody in the field."[67]

Another useful indication of the new importance of peer review is the decline of the "desk accept," a practice that was still common in the 1960s but had all but vanished a decade later. In a 1978 article for *The Sociological Quarterly*, Janice Beyer studied journals in physics, chemistry, sociology, and political science and found that the days when a significant percentage of manuscripts were approved without referees were gone. Beyer found that 94.6 percent of physics manuscripts and 99.7 percent of chemistry manuscripts were sent out to referees, and that 71 percent of the unrefereed physics manuscripts and 100 percent of the unrefereed chemistry manuscripts were desk-rejected. The number of unrefereed manuscripts was higher in the social sciences; 86.7 percent of sociology papers and 74.3 percent of political science papers were sent to referees. But the trend toward desk reject rather than desk accept was similarly pronounced; 87.7 percent of unrefereed sociology manuscripts and 97.2 percent of unrefereed political science manuscripts were rejected.[68]

Acceptance rates were also declining at many English-language scientific journals. Beyer found that the average rejection rate had crept up to

35 percent at physics journals and 29 percent at chemistry journals, significantly higher than the 15–20 percent averages of the 1950s and 1960s. In Lewin's recollection, the editorial team at *Cell* in the 1970s rejected roughly 80 percent of papers, often more; one purpose of the journal had been to call attention to the best and most interesting results in biology, and the editorial team set a high bar for "interesting."[69] For many scientists, high rejection rates were a feature rather than a bug. A high rejection rate seemed to signal higher standards to many employers and fellow scientists, and journals with outlier rejection rates were becoming sought-after places for publication. As scientific careers became more competitive in the 1980s, acceptance rates would continue to decline—and, as we shall see, concerns began to emerge about the possibility of scientists cutting corners or committing fraud to ensure their success.

7

INTO THE 1980s

"PEER REVIEW DEFINES WHAT IS SCIENCE"

In a 1986 essay for the journal *BioScience*, the science policy scholar Daryl Chubin reflected on what he called the "current round of scrutiny" surrounding peer review, by which he meant a collection of debates that covered everything from equity in research grants to concerns about scientific malpractice. Chubin argued that discussions about the purpose of peer review and its effectiveness were critical to the proper functioning of science. If scientists did not ask the crucial questions, he wrote, others would: "Without self-scrutiny of our quality control mechanisms, we court disaster by inviting hostile scrutiny." But even as he encouraged critique of peer review, Chubin's remarks revealed how deeply embedded the process had become in scientific institutions. If peer review decided whether something was "fundable or publishable," he said, the difference between "science" and "peer review" was so thin as to render discussion about the difference unconstructive. "Operationally," Chubin wrote, "peer review defines what is science."[1]

That same year, Eugene Garfield, creator of the Science Citation Index and the "impact factor" measure of journal performance, published the first in a four-part series of essays on refereeing in the natural sciences. Drawing from Harriet Zuckerman and Robert Merton's work, as well as essays by editors of scientific journals, Garfield offered an overview of refereeing's history along with an account of complaints about flaws in the system before concluding with his own views. Despite the frustrations, Garfield wrote, life was far better with peer review than without it. "No other activity is as fundamental to democratic scholarship," he wrote in the article's final section. "There is an Uncertainty Principle of Refereeing: The more we have of it, the less we like it—but the less we have of it, the more we miss it."[2]

It's worth pausing for a moment here to reflect on the timeline of the past few chapters. In 1973, the Office of Management and Budget seriously proposed eliminating the NIH's "peer review" system, which they saw as strange and indefensible. Two years later, witnesses at the NSF hearings framed grant peer review as the key to identifying quality science. By the mid-1980s, a prominent social scientist declared that peer review effectively "defines what is science," and the inventor of the impact factor saw peer review as "fundamental" to scholarship. This was a rapid shift in the discourse about peer review and its function in science—and it brought heavy pressures to bear upon the practice.

This chapter examines some of the consequences of peer review's new centrality to science in the US. During the 1980s stakeholders ascribed new professional and epistemological functions to peer review, such as deciding who deserved academic tenure and detecting fraudulent data. Universities across the US increasingly specified that only papers in refereed journals would count toward promotion. Accordingly, acceptance in refereed journals became an increasingly sought-after achievement for US-based researchers, especially as the number of faculty jobs declined following the economic contractions of the 1970s. Meanwhile, in discussions about fraud and other ethical concerns in the sciences, scientists argued that peer review would protect the integrity and trustworthiness of science. By the end of the decade, however, it seemed clear that peer review could at best provide imperfect protection against flawed or fraudulent research. As the expectations of peer review grew, so too did its opportunities to disappoint those who relied on it.

TENURE IN THE SCIENCES IN THE 1980s

As we saw in chapter 6, obtaining scientific research jobs in the US became significantly more difficult in the 1970s due to the decade's economic challenges. Hiring in the private sector was no longer as robust as it had been in the 1950s, and academic jobs became especially competitive. More and more newly minted PhDs were spending years in short-term "postdoctoral" appointments, moving between one- or two-year positions as they continued to apply for permanent jobs. Those who did eventually secure coveted assistant professorships then had six or seven years to demonstrate to their universities that they deserved tenure and permanent employment. Assistant professors who did not achieve tenure lost their jobs, making a tenure decision one of the most stressful points of a researcher's career. The bar for a tenure-track position was high, especially at the country's most prestigious universities. "To be considered for a tenure appointment," said one Columbia University memo in 1980, potential faculty members "must have achieved a very significant level of scholarly accomplishment and have extraordinary promise." Established scholars hoping for tenure "are required to be widely recognized as being among the leaders in the discipline."[3]

The expected CV of a tenured faculty member steadily grew longer over the course of the 1970s. In a 1980 article titled "Getting Tenure: Change and Continuity," the sociologist Lionel Lewis compared a cohort of tenure applications from 1967–1968 to tenure applications at the same unnamed university from 1977–1978. He found that in the natural sciences, 52 percent of the twenty-nine tenured applicants from the 1967–1968 cohort had published eleven or more articles; 10 percent had published twenty-five or more. Ten years later, there were only nineteen successful applicants from the sciences, and 69 percent of them had published eleven or more articles; 16 percent had published twenty-five or more. As the number of tenured faculty decreased, the average number of publications had crept up.[4] The types of articles that counted for Lewis's study were, specifically, "articles in refereed journals,"[5] reflecting universities' now-firm expectation that only peer-reviewed publications would count toward tenure and promotion. "Publishing in peer-reviewed journals" was "the heart of the tenure process," the psychologists Stephen Ceci and Douglas Peters wrote in 1982.[6] Ceci and Peters added that the peer review process also shaped academic careers through grants, since scientists were increasingly expected to bring grant money to their institutions to fund the work that would then form the basis of their peer-reviewed journal articles.[7]

To many in the academy, whether a piece of work was peer reviewed became a useful shorthand for whether it could be considered a meaningful research achievement. In a 1988 study of higher education merit pay, the economist W. Lee Hansen wrote that a candidate's research was the "easiest element to evaluate uniformly" because "peer review provides an effective mechanism for assessing the quality of research and is applicable to publications, grants, honors, and the like."[8] Universities and scholars alike increasingly referred to the tenure process itself as a "peer review" process, since departments solicited opinions both from the candidate's colleagues and from scholars outside the university on the significance of a candidate's scholarship.[9] Peer review now determined the fate of a scientist's career.

JOURNAL REFEREEING IN THE 1980s

In 1962, the University of Wisconsin chemist Howard Zimmerman (whom we met in chapter 3) declared that he would avoid publishing in US journals because he found refereeing intolerable. He seemed to see no professional downside to that choice; there were perfectly good European journals where editors still made the decisions. By 1990, however, the days in which a scientist could avoid peer-reviewed journals and still expect to be tenured or promoted at a US university were gone. The unrefereed scientific journal, already an endangered species in the US, went all but extinct in the English-speaking scientific publishing world. Competition for publication in top journals also became far more intense than it had been less than two decades earlier. In 1965, *Physical Review Letters*'s 50 percent rejection rate had been considered unusually stringent; by the early 1980s, top journals were rejecting up to 90 percent of submitted manuscripts.[10]

The personal papers of David Lazarus, who edited *Physical Review* from 1980 to 1990, clearly reflect the increased importance of refereeing to both *Physical Review* itself and to the authors who submitted their articles to that journal. In a 1984 letter to a rejected author, Lazarus wrote that the "desk accept" was no longer an option at *Physical Review*: "About

98% of papers are referred to external referees," and the rest were rejected outright.[11] "Papers are never accepted for publication without approval by one or more referees," he wrote in another letter. "No exception made for anyone!"[12]

In the 1960s, unhappy authors like Zimmerman had focused their frustration on journal editors, whom they thought were shirking their duties off to the cumbersome and unfair "referee system." In contrast, in the 1980s, the angry scientists who wrote to Lazarus to protest their rejections increasingly focused their ire on the referees themselves. Letters preserved in the Lazarus papers accuse referees of laziness, incompetence, and even outright malfeasance.[13] In response to one such letter, Lazarus encouraged his correspondent to think of referees not as adversaries but as fellow scientists. "Authors must understand that our referees, if imperfect (as are we all, sadly) are honest scientists who, at a minimum, are more expert than the average reader of an article in our journals," he argued. "If the referee finds an article confusing or unpersuasive, we must presume that an ordinary reader would be more confused or unconvinced."[14]

As scientific institutions and university hiring systems placed more emphasis on journal peer review, refereeing practices began to shift in subtle yet noticeable ways. By the early 1980s, the referee form at *Physical Review Letters* had grown from a short list of questions into a more substantial document that asked referees to comment not only on "the validity of the scientific content of the text" and "the apparent importance of the results" but also on a long list of formatting and writing considerations, including whether the title and abstract were informative, whether the tables and figures were "effectively presented," and whether there was "excessive use of jargon." The report should be "tactfully written," allowing the editor to communicate it directly to the author without redaction or rewriting.[15] Referee reports grew longer, as evidenced by several multi-page reports in the Lazarus papers—a far cry from Samuel Goudsmit's two-hundred-word reviews a few decades earlier.[16] At *Cell*, the editor Benjamin Lewin noticed a tendency for reviewers to become "picky," to want to "dot every I and cross every T" before they agreed a paper was ready for publication. Sometimes Lewin thought such pickiness came at a detriment to science; if a paper was "interesting enough and well-substantiated enough," he believed that "the appropriate thing is for it to be out there for everybody to see it." Reviewers' more stringent standards, he said, occasionally made it "more difficult to publish interesting things at the appropriate point."[17]

Reviewer anonymity also became a point of contention between journals and contributors, particularly among older contributors who were trained to expect a slightly more informal system. In 1984, the mathematical physicist E. J. Post wrote to Lazarus to suggest that referees should be identified.[18] When Lazarus wrote back to say that referee anonymity had been a policy for as long as he'd been an APS member, Post replied, "I suppose all of us grow up with different status quos . . . perhaps (and I may be older than you) there was more unofficial openness of refereeing in the past."[19] Post also claimed that referee identification was more common in mathematics, indicating that different scholarly disciplines still held diversity in reviewing practices.[20]

Lazarus's response to the suggestion of identifying reviewers was polite but unequivocal:

> Our use of referees for our journals is directed toward the end of getting expert advice to editors . . . Editors must make the ultimate decision, but this must be based on the best, most frank advice possible, with nothing held back to protect possible injured feelings. Thus, we must maintain anonymity of our referees, at all costs.[21]

The purpose of refereeing, for Lazarus, was not to give authors a satisfactory explanation for their rejection. Refereeing was for advising editors on whether a paper was suitable for publication, full stop. "For a paper to be published, it must be clearly transparent and comprehensible in its entirety to one or more experts," he told one unhappy author. If the expert referees could not follow the paper's argument or were not convinced by its conclusions, "we presume that it would not be of service to the profession to publish it for the benefit of more naive readers."[22] Lazarus cast refereeing as a gatekeeping measure, one that was necessary to select "comprehensible" papers that would be "of service to the profession."

PEER REVIEW AND DISCRIMINATION IN ACADEMIA

Lazarus seemed to view refereeing as a straightforward benefit to his journal and its readers, but other commentators were growing concerned that peer review might have a negative effect on science. Post-Mertonian scholars of science were taking a critical look at the inner workings of scientific institutions and highlighting discrepancies between rhetoric about the sciences and how scientific research worked in practice.[23] This new generation of scholars did not necessarily have a straightforward or favorable view of refereeing. In the 1990 book *Peerless Science*, for example, Chubin and his coauthor Edward Hackett argued that significant reforms were needed to justify peer review's "centrality" to the science–government relationship, including lessening the system's reliance on senior scientific figures and correcting its aversion to risk.[24] The Cornell sociologist Sheila Jasanoff studied peer review in government regulatory processes and expressed skepticism that the kind of "peer review" in use at journals or funding bodies would work for judging science in a policy environment.[25] In their provocative 1985 book *Lost at the Frontier*, the physicist Rustum Roy—whom we last encountered criticizing grant peer review at the 1975 NSF hearings—and the journalist Deborah Shapley argued that peer review was a "thoroughly vestigial ritual" that "saps the morale of the community" and "puts a premium on groupthink and straitjacketing of ideas."[26]

The concern that scientific institutions, including peer review, favored established people and ideas played a major role in controversies about discrimination in science during this period. In 1972, Title VII of the Civil Rights Act was amended to include employees in its protections against discrimination on the basis of race or sex. In the 1970s and 1980s, faculty members at several universities sued their institutions for employment discrimination. Lawsuits on the basis of racial discrimination had a low rate of success, and the high-profile case *Scott v. University of Delaware* may have had a chilling effect on racial discrimination cases in academia; the case was decided in favor of the university and suggested to many observers that racial discrimination cases would be extremely difficult to win.[27]

However, there were several highly visible victories for scientists who sued on the basis of sex discrimination during this period.[28] In one of the best-known cases, the chemist and postdoctoral fellow Shyamala Rajender sued the University of Minnesota in 1973 after her department chair told her that appointing her to a tenure-track job would ruin the "collegiality" of the all-male department.[29] Rajender ultimately won her case, although by then she had already left academia to become a patent lawyer. The case nevertheless left a mark on the University of Minnesota, which was required to create and abide by gender-neutral hiring and promotion procedures.[30]

Another widely covered case was that of Jennifer Harrison, a mathematician at the University of California at Berkeley whose case for tenure was denied in 1986. Harrison sued her university, alleging that both the initial decision and the appeals process indicated that the university was biased against women. "I had at least as many papers as three men tenured during the time I was at Berkeley," Harrison told *Science* in 1991.[31] Harrison and her supporters accused the department vice chairman Marc Rieffel of putting his thumb on the scale of the review process by soliciting more letters than usual from fellow mathematicians, implying that Rieffel had been fishing for at least one negative letter. After a protracted legal battle, the case was settled out of court and a new review panel, appointed as a condition of the settlement, voted to tenure her.[32]

Harrison and Rajender were both researchers in overwhelmingly male-dominated fields, a fact that helped their legal cases. At the time Harrison was denied tenure, the country's top ten mathematics departments in the country (including Berkeley) had only four tenured female faculty members between them. Rajender's department at the University of Minnesota had no women in tenured or tenure-track positions; out of sixty-nine prospective faculty candidates interviewed between 1968 and 1980, only one was a woman. At trial, Rajender's lawyers presented an evaluation sheet for the sole female candidate written by the chemistry professor Edward Leete. "I have to state that she would have problems because she is a woman," Leete wrote. "I guess I am a male chauvinist pig."[33]

Supporters of plaintiffs like Rajender and Harrison argued that their cases proved that science was not living up to its lofty meritocratic ideals. Several prominent male spokesmen for both universities and US scientific societies, however, insisted that rhetoric and practice were perfectly aligned. These spokesmen insisted that the forms of "peer review" involved in hiring and promotion were purely meritocratic. In their telling, women who did not obtain academic jobs were simply not qualified for the positions. Philip Handler, president of the National Academy of Sciences, decried government attempts to diversify state university faculties, arguing that the government was trying to "achieve social goals irrelevant to the research enterprise per se, most notably 'affirmative action' in appointment to the faculty of women, blacks, and other minorities." He declared that "nothing can so damage the future of a university as an appointment to the faculty of anyone less than the best."[34] In a public statement on Rajender's case, Sheldon Steinback, an attorney for the American Council on Education, deemed the ruling "a disaster for higher education" and "part of a pattern of erosion of the peer review process."[35]

The courts, too, occasionally suggested that legal intervention in academic hiring risked undermining expert judgment. In one case, involving a biochemist named Sharon Johnson who had been denied tenure at the University of Pittsburgh, the judge stated in his ruling that the courts should not "take over the matter of promotion and tenure . . . we must leave such decisions to the Ph.D.'s in academia."[36] In short, defenders of academic hiring practices characterized both discrimination lawsuits and efforts to diversify academia as an attack on peer review. The scientists alleging discrimination, defenders claimed, were attempting to do an end run around peer review, using the courts instead of their research and publication record to obtain positions.

A series of psychology studies, however, seemed to confirm worries that the peer review system might be biased against women. An influential 1968 article by the psychologist Philip Goldberg had found that female college students judged articles in fields from law to city planning more harshly if a woman was listed as the author.[37] Revisiting Goldberg's findings in the 1980s, several studies found that readers of all genders were more likely to rate an article favorably if it was attributed to a male author. In one study, the psychologists Michele Paludi and William Bauer gave their subjects articles attributed to "Joan T.," "John T.," or "J.T.," and asked them to read each article and rate its quality. "Each of the three articles [used in the study] was rated least favorably when it was supposedly written by a female author," they reported.[38] A study by psychologist Colleen Ward found "a marked tendency for males to denigrate the female author" over "status and competence in the field" in their comments on an academic article.[39] In an update to Merton's work, the historian of science Margaret Rossiter pointed out that women were often left out of the reward system of science because their contributions were frequently attributed to male collaborators.[40]

Notably, studies about protected class discrimination from this period were focused almost entirely on sex and generally did not consider other factors such as race in their analysis. One non-sex factor that did receive scholarly attention, however, was the relative prestige of a scientist or an institution. In one of the most cited studies from this period, the psychologists Stephen Ceci and Douglas Peters submitted twelve previously published articles to the high-status psychology journals that had initially accepted them. They changed the names of the authors; they also changed the institutional affiliations to "fictitious ones without meaning or status in psychology," such as the "Tri-Valley Center for Human Potential."[41] Three of the articles were rejected on the grounds that they were plagiarized—the ideal outcome, given that the articles were copied word-for-word from previously published articles. Of the remaining nine, however, eight were rejected on the grounds of poor fit or insufficient quality. The most likely explanation, Peters and Ceci argued, was that "institutional affiliation or professional status can in fact bias peer review."[42] Similarly, the sociologist Michael Gordon found that referees in the physical sciences were far less likely to rate a paper as "good" if it came from a "minor university."[43]

In response to some of these concerns, some journals in the social sciences began to experiment with "double-blind" refereeing, in which the referees were not given the author's names.[44] Proponents of anonymous-author refereeing often argued that it gave a fairer assessment of work by junior scholars, minority racial groups, and women.[45] An eye-opening study by the classicist Mary Lefkowitz seemed to confirm that anonymous-author refereeing might prevent some kinds of discrimination. In 1974, Lefkowitz studied the fates of submissions to the American Philological Association (APA) annual meeting and found that removing the authors' names nearly doubled the number of accepted abstracts submitted by women. Both the APA and the Modern Language Association (MLA) adopted anonymous-author refereeing for their journals and conferences following Lefkowitz's findings.[46] The American Psychological Association, the American Sociological Association, and the American Political Science Association all adopted anonymous-author review by the mid-1970s.

Journals in the natural sciences, however, largely continued (and continue) including authors' names with their manuscripts. Many natural scientists saw that system as a bulwark against "cranks" who lacked the proper training to make scientific contributions.[47] There was also a belief among some journal editors that taking authors' names off an article deprived reviewers of relevant information. As the *Cell* editor Lewin would later explain it, in the days before online appendices, most papers had to refer to unpublished data—which he believed made the author's reputation an important part of judging the paper:

In that period it would be a rare paper that didn't have some supporting points stated as unpublished data. You want to establish a reputation in science [so] that when you say I have unpublished data showing X, people will believe you. . . . You establish a certain reputation in science and that reputation has an influence on the people who are reviewing your latest paper. That's the way the system works.[48]

Similar reasoning about the importance of the applicant's background and track record was used for grant peer review, where anonymous-author review was not seriously proposed in any field. The psychologists Cannie Stark-Adamec and Robert Adamec criticized grant peer review specifically as riddled with "problems and injustices," due in large part to the fact that applicants' names were known to reviewers. "A reviewer can say the most personal, insulting, irrelevant things about you with impunity," they wrote. "You have no recourse. It is not like a review of a manuscript where you can send a rebuttal to the Editor of the journal."[49] Stark-Adamec and Adamec argued that grant peer review was an "old boy's network" that disadvantaged female applicants and hindered their chances of promotion and tenure in the sciences. But many scientists continued to assert that the track record of a principal investigator and the facilities available to the project were necessary information to evaluate a proposal's chances of success. Observers had a difficult time conceiving of a grant application that did not include its proposers' names, institutions, and CVs.

BIOMEDICAL FRAUD IN THE 1980s

The integrity of grant review was a particularly sensitive subject due to a series of scandals involving NIH-funded projects in the 1980s. As we saw in chapters 5 and 6, during the 1970s scientists convincingly argued that the only way to ascertain the quality of a piece of research was to subject it to review by one's peers. Scientists, legislators, and members of the public took this argument to a logical but perhaps unintended conclusion: that peer review was capable of detecting *bad* research, including work based on falsified data. Several high-profile scientific scandals in the 1980s, however, called peer review's effectiveness as a fraud detection mechanism into question.

In the early spring of 1981, Al Gore Jr., a second-term Democratic congressman from Tennessee, chaired two days of hearings before the Committee on Science and Technology's Subcommittee on Investigations and Oversight. The impetus for the hearings was a small but unsettling group of recent scientific frauds, all involving biomedical journals or NIH funds.[50] At Boston University, a physician had fabricated results for a cancer research study; at Yale, a researcher had falsified data for a study on Hodgkin's lymphoma; in Texas, a researcher had plagiarized dozens of articles and published them with names of coauthors who had never worked with him. Gore opened the hearings by observing that the US Federal Government would spend $35 billion on research and development in 1981, with $3.4 billion going to the NIH. And yet, he said, the recent high-profile incidences of fraud suggested that Congress needed to examine how grants were awarded and how fraud was reported. "Is science really self-correcting?" Gore asked in his opening statement. "Is the peer review process working adequately? What responsibilities do referees and journal editors have to insure [sic] the integrity of data?"[51]

The scientists called to testify at the 1981 hearings argued that the scrutiny of one's fellow scientists was the best way to ensure that funds on biomedical research were not spent on fraudulent projects. "I have yet to hear of a genius who is possessed of such clairvoyance that he or she falsified important data that stood the test of time and the scrutiny of scientific peers," Handler, president of the National Academy of Sciences, said in his testimony. "In the hot cauldron of the operation of peer review in the scientific world, such data cannot long survive undetected."[52] The associate NIH director William Raub argued that "the first critical step is peer review of proposed research" and that scientific journals could be trusted as sources of information because "new scientific findings enter the literature through an equally rigorous peer review process."[53] Raub admitted that peer review was "imperfect" as a detector of fraud and that "it is appropriate that we not rely upon it totally," but he defended it as a critical step that would root out almost all cases.[54]

Legislators at the 1981 Gore hearings, however, were more skeptical than their 1975 counterparts at the NSF hearings had been. Gore repeatedly raised questions about whether the peer review process was actually capable of detecting fraud, as did the Pennsylvania congressman Robert Walker, the ranking Republican member of the subcommittee. Walker pointedly remarked that many of the fraudulent papers discussed in the hearings had been "submitted to the peer review process on a couple of occasions and passed through that process," and he argued that "the peer review process has no real validity as a safeguard."[55] Walker felt that scientists were too quick to assume each other's honesty. "It seems to me there is a large degree of trust built within the peer review system, and maybe that is as it should be, but it is also the kind of thing which causes very, very noticeable failures," he said on the second day of testimony.[56]

Those kinds of criticisms were aired outside the halls of Congress as well. In the spring of 1982, the science journalists William Broad and Nicholas Wade published a book titled *Betrayers of the Truth*. Broad was a relatively young writer just five years out of journalism school; Wade was an established figure in the science journalism world who had been an editor at *Science* for the past decade and would join *The New York Times* that year. The two writers focused on the same series of scandals in biomedical research that had led to the Gore hearings a year earlier. Broad and Wade felt strongly that scientists were not taking the potential consequences of such episodes seriously enough. Public trust in science was at stake, the pair said, and the scientific community was too quick to find comfort in the assumption that flaws or problems would be easily detected by fellow scientists.[57]

Betrayers of Truth argued that scientists were as self-interested and motivated by personal advancement as any other career group, and that the structure of modern science made fraud both increasingly difficult to detect and more likely to occur. Scientists, said Broad and Wade, relied on three self-policing mechanisms to guard against scientific fraud: peer review of grants, refereeing of journals, and replication of experiments.

All three mechanisms, however, had failed to prevent the dozens of cases of misconduct covered in the book, cases the authors said went all the way back to the astronomer Claudius Ptolemy in the second century AD.[58] Prone to elitism and more than a drop of randomness, neither peer review nor refereeing were capable of detecting either flawed or fraudulent science. "Peer review and refereeing are at best coarse screens, not the infallible, fine discrimination systems that scientists often portray them to be," Broad and Wade declared.[59]

Concerns about bad research—and whether peer review could prevent its publication and funding—only intensified over the course of the decade following two high-profile scandals, one involving a physician named John Darsee and one involving the biologists David Baltimore and Thereza Imanishi-Kari. On the surface, the episodes had little in common. Darsee's story was a fairly clear-cut case of data fabrication, while the Baltimore Affair (as it came to be called) was a layered, ambiguous, and painful controversy without a clear resolution. Both episodes, however, seemed to call into question peer review's ability to detect either deliberate fraud or work that was simply flawed.

When he arrived at Harvard Medical School (HMS) as a research fellow in 1979, thirty-one-year-old John Darsee seemed destined for academic superstardom. He had graduated from medical school in 1974 and quickly made his mark at Emory University's Grady Memorial Hospital, where a senior colleague described him as "one of the most remarkable young men in American medicine."[60] He built his reputation not only on his clinical work but on his research. Between 1974 and 1981, Darsee published over ten articles in top medical journals as the first or last author and was a coauthor on dozens more, an extremely impressive publication record by the standards of the day.[61] By 1981 HMS was moving to hire him on its faculty at the urging of the legendary cardiologist Eugene Braunwald, and he had been given a pivotal role on a large NIH-funded project assessing the effectiveness of different heart attack drugs.

Darsee's career would collapse just a year later under allegations of serial fraud that called nearly every paper he'd ever published into question. Colleagues of Darsee approached an HMS professor with suspicions that Darsee was fudging some of his data on a joint project.[62] An initial investigation indicated that Darsee was claiming to have generated weeks' worth of data in the span of days or hours. The effort to place him on Harvard's faculty came to a screeching halt, and Darsee's NIH-funded fellowship was terminated, but little action was taken beyond that. Darsee's supervisors apparently believed his actions were the result of a brief lapse in judgment and that the young researcher had learned his lesson. Significantly, HMS did not inform the NIH of what they had found. However, when the five institutions involved in the NIH-funded cardiac drug study began comparing their preliminary data in October 1981, it became clear that Darsee's data was a major outlier, and his supervisors began to suspect that his data fabrication extended to that work as well. Subsequent investigations by Harvard, Emory, and the NIH each found that Darsee had published falsified results.[63] At that point, Darsee left academic medicine with little resistance or fanfare, an outcome that seemed appropriate to most observers.

The Baltimore Affair, a complicated scandal that had its origins in an NIH-funded collaboration between the Nobel Prize–winning immunologist David Baltimore and the biologist Thereza Imanishi-Kari, was a far longer and messier episode.[64] In a 1986 paper in *Cell*, Imanishi-Kari (the lead author), Baltimore, and their coauthors argued that transgenes—segments of DNA that had been transferred from one organism into another—could change antibody production in mice.[65] The finding seemed to indicate that transgenes could one day be a key to strengthening immune systems, particularly in immunocompromised patients. But a postdoc named Margot O'Toole, hired to Imanishi-Kari's laboratory in June 1985, struggled for months to replicate the results her mentor had reported in *Cell*.[66] Imanishi-Kari was dismissive and harsh when O'Toole reported her difficulties, and O'Toole eventually came to believe that Imanishi-Kari had intentionally published false data. O'Toole took her concerns to MIT and Tufts University, the two institutions that had employed Imanishi-Kari during the work. Both universities heard O'Toole out, but they did not find reason to dismiss Imanishi-Kari.

O'Toole's concerns, however, soon found a more sympathetic audience: Walter Stewart and Ned Feder, two NIH employees who had developed a reputation for fraudbusting. The pair had been involved in the Darsee affair as well. As Stewart would tell the story in a 1989 interview with *Omni* magazine, Darsee's case caught his attention when his and

Feder's alma mater, Harvard, sent alumni an account of the university's investigation. The pair was immediately skeptical of the university's claim that they had detected just three instances of demonstrable fraud.[67] They decided to conduct their own analysis of Darsee's papers and found that Darsee's work "contained a very large number of obvious errors." Most damning was the leading figure in a *New England Journal of Medicine* paper, which

> introduced a human pedigree [genealogy] showing the inheritance of a new disease that the paper described. . . . I noticed the paper included a seventeen-year-old with this disease. He was listed as having four children, including an eight-year-old daughter.[68]

Stewart laid the blame for the Darsee case squarely at the feet of Darsee's referees. He said that he felt the coauthors "have a responsibility to ensure the accuracy and truthfulness of the entire paper," but, as he told *Omni*, "we thought the journal's referees should have caught this kind of thing."[69] Stewart and Feder now began the first of several investigations into Imanishi-Kari's work.

Meanwhile, the Michigan congressman John Dingell became interested in the case and decided to hold a series of hearings on the paper. The government investigations into the *Cell* paper would stretch out for a full decade, inflicting serious consequences on nearly everyone involved. Imanishi-Kari ended O'Toole's postdoc, and O'Toole and her husband lost their house.[70] During the Dingell investigation, Imanishi-Kari's laboratory notebooks were seized and subjected to a Secret Service analysis to see if she had deliberately written over sections of data. Although Baltimore himself had not personally conducted any of the challenged lab work, his status as a Nobel Prize winner made him the most notable name in the controversy. Baltimore resigned as president of Rockefeller University, although he kept his faculty position there. He also asked *Cell* to retract the paper. The affair did not truly draw to a close until 1996, when a Department of Health and Human Services appeals panel exonerated Imanishi-Kari and her *Cell* coauthors of deliberate wrongdoing.[71]

On the surface, although the Baltimore and Darsee cases both involved the biomedical sciences, they had little in common. The Darsee story was about one bad actor who had fabricated data and suffered professional consequences for it. The Baltimore case was far more ambiguous. The scientific community was largely sympathetic to O'Toole, who seemed to honestly believe she had detected wrongdoing and had an obligation to report it, but most observers felt that the more likely explanation for difficulties replicating the *Cell* data was flawed or sloppy initial work—a valid reason for criticism but not something that ought to end a scientific career. Despite their differences, both episodes drew attention to the peer review system, with each seeming to call into question peer review's ability to perform the epistemological tasks it had been assigned. "The most important issue" raised by the Darsee case, said Emory's alumni magazine, "is whether the system of peer review upon which science so heavily depends has the necessary safeguards against researchers who unethically manipulate their findings."[72] In an interview with the *Washington Post* just before his testimony at the Dingell hearings, Stewart argued that peer review was unequal to the task of detecting fraud: "The peer review system does not pick up fabricated work. You can get away with murder for a pretty long time."[73]

In response to these criticisms, scientists doubled down, holding fast to the claim that peer review could prevent low-quality or fraudulent contributions to the literature. The combative tone of the Dingell hearings led most scientists to feel that Imanishi-Kari and Baltimore were the subjects of unfair persecution, and much of the commentary from the scientific community argued that such hearings were unnecessary when scientists had peer review as a check on bad work. "We continue to build on our traditions of trust and autonomy in promoting excellence by example and peer review rather than by legal procedures," Bernard Davis, a retired Harvard professor of physiology, wrote in an editorial for the *Los Angeles Times*.[74] Declaring his "deep respect for the peer review processes that define the funding of biomedical research in this country," the biologist Robert Pollack wrote in *The New York Times* that the Dingell investigation was attacking "the legitimacy of the scientific method itself."[75] No one suggested that the fact that the *Cell* paper had passed through peer review put it above question, but scientists did express their strong belief that science could and should police itself—and that the centrality of peer review to both grant-making and publishing was one of the major reasons that they could be trusted to do so.

To many observers, however, scientists' confidence seemed misplaced. There seemed to be little evidence that refereeing could detect fraud reliably. A 1983 NIH internal review of scientific misconduct investigations listed forty-eight recent cases, along with how they were reported; just three of those cases were caught by other scientists during peer review. It was far more common for the scientists' own institutions to report them or for a "third party informant" to be listed as the source of the report.[76] The *Cell* editor Lewin remarked that the causes of fraud or serious error that he was aware of as editor were usually uncovered because whistleblowers came forward, not because a referee spotted something amiss.[77]

More cautious voices warned that the fraud scandals of the 1980s could reverse some of the support for scientific autonomy and peer review that scientists had secured barely a decade before. In an article titled "Baltimore's Legacy: Concern About Oversight of Scientists," the *Boston Globe* reporter Richard Saltus wrote that observers increasingly felt that "Congressional attention to the issue of scientific misconduct is not going to go away" and that "the science establishment has not yet demonstrated that it can police itself."[78] An article in *Science* quoted Stephen Lock, the editor of the *British Medical Journal*, warning that "if we don't put our house in order, the chaps on Capitol Hill and the House of Commons will."[79]

PEER REVIEW GOING GLOBAL

Lock's comments were made at a notable venue: the First International Congress on Peer Review in Biomedical Publication, held in Chicago in May 1989. As the organizer Drummond Rennie would recount thirty years later, the impetus for the meeting was that "spectacular cases of what came to be called research misconduct" had "forced medical journal editors to think hard about the processes that had allowed fraudulent research to be published despite their systems for expert peer review."[80] The American Medical Association, publisher of the prominent *Journal of the American Medical Association* (*JAMA*), agreed to fund a conference about peer review, and nearly three hundred attendees gathered in Chicago to hear papers by editors and social scientists about different refereeing systems, including some at funding bodies.

The term "International" in the title of the conference signaled an important development: Peer review was starting to go global. The use of referees outside the US, of course, was nothing remarkable; journal referees were a European invention, as we saw in chapter 1. Funding bodies outside the US had, by this point, also been using referees for some time. For example, Canada's Medical Research Council, founded in 1960, brought in volunteer experts to assess proposals early in its history.[81] In 1977, the West German government funding body Deutsche Forschungsgemeinschaft (DFG) created an "open call" program with formal application guidelines that required a grant proposal, with the explicit intention to subject those proposals to peer evaluation.[82] As acceptance rates declined in the 1980s, DFG increasingly emphasized that decisions were being made based on peer review. As one DFG document told applicants, "We believe that a good researcher is always capable of identifying and evaluating good research."[83]

So non-US referees were nothing new—but the idea that "peer review" was crucial to properly conducted science *was* new as of the 1970s, and the program of the First International Congress on Peer Review shows that the idea was starting to be adopted outside US borders. The majority of papers and posters at the congress, 34 out of 45, came from US-based researchers or research teams. But several British biomedical journal editors participated, as did editors and researchers from the Netherlands, Canada, and India. Adoption of peer review had not been uniform; the biostatistician B. L. Verma's work on journals in India, for example, found that a significant number of Indian biomedical journal editors eschewed external review "on the grounds of cost as well as man-hours involved."[84] But other abstracts made it clear that "peer review" was becoming more central to scientific publishing all over the world. "I believe that publication is central to science and that the peer review process is central to scientific publication," Rennie wrote in the "Welcome!" section of the program. "Your presence here in Chicago suggests that you agree."[85]

The rise of English as the dominant language of scientific communication was almost certainly one reason why the rhetoric about peer review's centrality had spread. At the turn of the twentieth century, back when refereeing was a curious practice confined mostly to Anglophone scientific societies, the most common second language for scientists to learn as part of their training was not English but German—but an international backlash following the World Wars made German unacceptable to wide swaths of international scientists. Meanwhile, as the US government poured money and resources into scientific research and scientific institutions after World War II, the US scientific community grew in size and influence—and the US, by and large, was a country that had resisted training young people in foreign languages.[86] To read the US literature, scientists all over the globe had to learn English, and if scientists wanted their articles read by their US colleagues, they had to write them in English too.

European for-profit science publishing firms such as Pergamon Press, Elsevier, and Springer began publishing the majority of their journals in English as early as the 1950s. That development is especially notable for Elsevier, a Dutch company, and Springer, which is German.[87] A paper at the First International Congress about the peer review experiences of Dutch PhD students explicitly stated, "Dutch scientists often seek wider publicity for their work by publishing in English-language journals of standing abroad."[88] By 1980, nearly 75 percent of the worldwide scientific literature was published in English.[89]

But as "peer-reviewed research" became the criterion for hiring, promotion, and tenure in the US, merely publishing in an English-language scientific journal was no longer enough. For an article to count toward career advancement in the US, it had to be peer reviewed. International journal editors were well aware of this dynamic, and they were concerned about their journals' ability to attract US-based authors and readers unless they adopted peer review systems. David Davies, the Welsh geophysicist who edited *Nature* from 1973 to 1980, had been working at MIT when he accepted the position at the helm of the journal. Davies's predecessor John Maddox had been comfortable accepting articles without referee reports, but Davies would later recall that he and his American colleagues were alarmed by *Nature*'s unsystematic attitude toward refereeing. He felt that "getting the refereeing system beyond reproach" was his most important task as editor.[90]

Still, there were holdouts. By the late 1970s, the editor of the British medical journal *The Lancet* worried that its articles might not be taken seriously in the US unless the journal began systematically consulting referees rather than allowing the editor to accept papers on his own authority.[91] In the end, however, *The Lancet* stayed its course. In a 1989 editorial, *The Lancet* complained that "in the United States far too much is being asked of peer review" and proudly assured readers that at *The Lancet*, "reviewers are advisers not decision makers."[92]

CONCLUSION

"It's called peer review," Jon Turney, the science editor of the London *Times Higher Education Supplement* wrote in a 1990 essay. "For a long time it worked quite well. The tradition of asking scientists of repute to judge the quality of others' work goes back to the birth of the *Philosophical Transactions*." But, said Turney, "most scientists agree that peer review is a deeply flawed method."[93] The rising rejection rate for grants, the increasing labor burden on reviewers, and the perception that peer review was slow to recognize innovation all contributed to the frustration Turney saw with the system. But, said Turney, recent surveys had found that most scientists believed that science needed peer review. "That faith is still with us," Turney concluded. "Peer review is like democracy; highly imperfect in many ways, but better than the alternatives."[94]

Turney's remarks nicely summarize the arc of this chapter. In the 1980s, scientists grew increasingly concerned that peer review was not "working" the way it was supposed to. Different observers complained that peer review did not detect fraud or sloppy research, that it discriminated against minoritized groups, or that it favored big names and established ideas. Episodes like the Darsee scandal and the Baltimore affair offered compelling evidence that peer review was not performing some of the tasks it had been assigned. Yet none of those concerns resulted in major changes to peer review systems, which were now thoroughly embedded in US scientific institutions and were growing more influential outside the US as well. Despite its "highly imperfect" nature, peer review seemed to be here to stay. Not only that: To return to Chubin's words, peer review "defined what is science."

I want to take a moment to reflect on Chubin's claim as we move toward the conclusion. At the risk of nitpicking, I would argue that it was not "peer review" itself that defined science. Peer review is a system; it has no independent agency and can take no action without human involvement. It was, more specifically, the *people* involved in peer review who defined science, including the editors and funding body officials who made decisions based on referee reports, the university committee members who read tenure letters, and most importantly, the referees whose expertise was the justification for using peer review in the first place. By embracing peer review as a key to publishing, funding, and tenure, scientists ensured that their judgment would matter in those spaces. Even if there were some flaws to work out, there seems to have been a general agreement that the idea behind peer review was the right one and that the peer review system placed judgments about science in the hands of the correct people: scientists themselves.

Despite some skeptics like Broad and Wade, who raised concerns about cronyism or conflicts of interest, the idea that experts in the field should be the ones to judge the value of a proposal or paper was largely uncontroversial at the close of the twentieth century. By 1990, US scientists no longer held the lofty social or cultural place that they had in the 1950s, but the faith in the value of scientific expertise was still deep enough and broad enough to sustain a belief in peer review's essential validity. As I revise this chapter in summer 2025, however, US science finds itself in a far different historical moment—one that may well be a turning point for many of the institutions I have written about in this book, as well as the peer review system itself.

CONCLUSION

In May 2025, I received three extremely helpful referee reports on this book, all of which made valuable suggestions on how to improve the manuscript. The feedback on my original conclusion confirmed something I had increasingly suspected: that I would have to rewrite it from scratch, in large part because the situation for science and scientists in the US had changed dramatically in the months since I sent my draft manuscript off to the MIT Press. On January 20, 2025, Donald Trump was inaugurated for a second term as president of the United States, and ever since, the Trump administration has been dismantling many of the systems I wrote about in this book. Events have been developing so rapidly that I am sure what I write here will be dated by the time this book reaches publication; every day it seems that there's a new headline, a new executive order, a new judge handing down a new temporary injunction. But I'll highlight a few developments from the past months that are particularly relevant to our topic.

Shortly after Trump's inauguration, peer review panels at the NIH and NSF were suspended pending an examination of whether their work fit the president's agenda.[1] The US government began canceling research grants across the country.[2] In July, the administration dismissed dozens of scientists from NIH review panels.[3] In August, the president directed federal funding agencies to "designate a senior appointee" who would "review discretionary grants to ensure that they are consistent with agency priorities and the national interest;"[4] this was followed by a December memorandum from the OMB ordering agency heads to "put political appointees in control of [the agency's] grant process to deliver results."[5] After leaving the National Science Board, the sociologist Alondra Nelson reported that she resigned because Trump's new Department of Government Efficiency had been given "the authority to give thumbs up or down to [NSF] grant applications which had been systematically vetted by layers of subject matter experts."[6] The administration is proposing budget cuts that would halve scientific funding at the NIH and NSF.[7] There have been mass purges of federal employees, including scientific researchers.[8] An interim US attorney for the District of Columbia sent a letter to the editors of scientific journals demanding that they explain how they guaranteed "acceptance of competing viewpoints."[9] The administration mass-canceled subscriptions to Springer Nature journals for federal research institutes.[10] The secretary of Health and Human Services (HHS), Robert F. Kennedy Jr., has indicated that he may forbid NIH scientists from publishing in independent medical journals, instead requiring them to publish all of their work in-house.[11]

I had thought I was writing a biography of peer review, calling attention to the way Cold War US federal funding shaped the scientific community's language about and reliance on the practice. Lately, however, it feels more like I've written an obituary.

PEER REVIEW WORLDWIDE

The idea that the Trump administration will kill peer review is probably a bit too dramatic. For one thing, it's notable that Trump, Kennedy, and those supporting the administration's gutting of federally funded research insist that they are doing so in *defense* of good science, including peer review. A recent Trump executive order on "Restoring Gold Standard Science" declared that "unbiased peer review must be maintained."[12] Trump's science advisor Michael Kratsios has declared that the administration will make certain that "taxpayer-funded science" has been "subject to unbiased peer review."[13] The vaccine-skeptical journals favored by the new HHS leadership, such as *Journal of the Academy of Public Health* and *Science, Public Health Policy and the Law*, proudly proclaim their commitment to "rigorous peer review."[14] The idea that science must undergo "peer review" still has a great deal of rhetorical power—at least for now.

More significantly, peer review is no longer the unusual American obsession that it was in the 1960s and 1970s. It is now a practice in scientific institutions worldwide. I brought the main narrative of *In Referees We Trust?* to a close around 1990 because a full account of how peer review went global would require its own book, but a few more thoughts on that spread are worth offering here. In chapter 7 we saw that the use of English as the global language of science publishing exerted important influence over the adoption of peer review at journals. By the new millennium, peer review had been incorporated at nearly every English-language scientific journal—and those English-language journals were an even larger share of the market than they had been during the Cold War. Between 1980 and 1996, the percentage of worldwide scientific articles published in English went from an already-high 74.6 percent to an overwhelming 90.7 percent.[15] The term "peer review" and the rhetoric surrounding it also became increasingly global. At the Royal Society of London, for example, internal communications began using the term "peer review" in the 1990s, and by the early 2000s, the society was boasting that it had in fact invented the peer-reviewed journal.[16]

Although peer review rose to a central place in US science first, I want to caution against assuming a "diffusionist" story in which peer review inevitably spread from the US to other countries. Existing accounts of peer review adoption outside the US all make one thing clear: Such adoptions took place in response to the perceived needs of different national and linguistic scientific communities. When the USSR dissolved, for example, Soviet scientists took the opportunity to decentralize science by arguing that funding applications should be peer reviewed rather than granted by the directors of research institutes.[17] At a moment of transition, Soviet scientists reached for peer review not because they felt they had to follow the US example but because peer review suited their own goals.

Today scholars from around the world assume their work will go through the channels of peer review to receive funding or be published in an academic journal. The Brazilian *Anais da Academia Brasileira de*

Ciências informs aspiring authors that "articles will be evaluated by at least two reviewers" who have "proven scientific production."[18] The Indian Academy of Sciences's *Journal of Genetics*, the *Journal of the Korean Physical Society*, and the modern *Comptes Rendus* of the Academié des Sciences all assure potential authors that they are peer reviewed as well.[19] In worldwide grants, the British Medical Research Council website has a page devoted to the details of its "external peer review process,"[20] the DFG assures applicants that it consults over seventeen thousand external referees per year,[21] and the Japanese Society for the Promotion of Science informs potential applicants that funded projects "are selected using a peer-review screening process (screening by multiple researchers whose field of specialization is close to that of the applicant)."[22] It seems likely that, whatever happens in the US, peer review will survive elsewhere in the world, especially since the proportion of papers written by scientists in the US is already in relative decline due to the rapid growth of papers by scientists working in other countries.[23]

FRUSTRATIONS WITH PEER REVIEW

Peer review, then, is now a truly global practice—but, just as we saw in chapter 7, its expansion has been accompanied by growing concerns and complaints about whether it is working the way researchers think it should. In 2011, Great Britain's House of Commons commissioned a report on the state of peer review and concluded that while peer review "is crucial to the reputation and reliability of scientific research," many scientists believe the system stifles progress, is often biased, and that "there is little solid evidence on its efficacy."[24] Studies have found that referees are unlikely to spot errors in papers, even major ones.[25] There are ongoing concerns that peer review stops truly innovative work from being funded or published.[26] There are also rising concerns that the amount of time scientists dedicate to peer review is unsustainable and that shortages of reviewers are leading to long delays between submission and publication, an ironic development given that refereeing used to be seen as the *solution* to labor problems.[27]

The widely read website Retraction Watch often discusses problems in the peer review system, ranging from reviewers who approved obviously sloppy or flawed research to scientists who game the system to review their papers themselves.[28] In a 2024 interview, the Retraction Watch cofounder Ivan Oransky gave a blunt assessment of the state of academic peer review: "Things are pretty bad and they have been bad for some time."[29] Some scientists wonder if systematic peer review is still the best method of evaluating scientific research and have argued that peer review could be eliminated with little cost to the quality of the scientific literature, pointing to examples like the successful preprint server arXiv.org to argue that scientific communication does not need prepublication review.[30] The psychologist Adam Mastroianni has called peer review "the most massive experiment ever run," but he argues that the system has "failed" and needs to be discarded. "I don't know what the future of science looks like," he writes, but "it'll be a lot better than what we've been doing for the past sixty years."[31]

But there is a reason why peer review persists despite the many complaints about it. By now, peer review is so enmeshed in research institutions that major changes to it would be extraordinarily disruptive. The increases in peer review's status and rhetorical power that I described in chapters 6 and 7 are still with us and have only grown in the twenty-first century. Scholars in disciplines from English literature to computer science rely on peer review processes to choose between papers and grants. Academic promotion and tenure still depend on a record of peer-reviewed publications; junior scientists who choose non-peer-reviewed journals would be risking their careers. Government funding bodies that try to change or eliminate their peer review systems would be met with an outcry from researchers, as we are seeing now in the US.[32]

Furthermore, the available data indicates that despite their frustrations with peer review, the majority of researchers still believe that peer review contributes something valuable to science. A 2015 report commissioned by the Publishing Research Consortium, a group of for-profit and nonprofit scholarly publishers, found that 74 percent of respondents believed that peer review improves the quality of published papers and 75 percent of respondents either agreed or strongly agreed that "scholarly communication is greatly helped by peer review of published journal papers." The belief that peer review ought to be able to detect fraud has also proven remarkably durable. A full 82 percent of respondents indicated that they thought peer review should detect fraud, although only 41 percent said peer review was currently doing that.[33]

As I admitted at the beginning of this book, I do not have straightforward answers about how to "fix" peer review. History is not a predictive discipline; studying how refereeing changed in the 1970s does not give us a recipe for changing it again fifty years later in response to an entirely different set of conditions and concerns. But I do think understanding the past can inform where we go in the future and how we get there. As researchers consider the future of peer review, I believe they will need to take two central and related questions into account, questions whose answers have shaped refereeing at each step in this book. First, what is peer review for? And second, to whom is an institution that employs peer review accountable?

If peer review's only function was to prevent flawed or fraudulent research from reaching circulation, and if our scientific institutions were only accountable to scientists, it might be reasonable to get rid of it. Bluntly, peer review doesn't detect fraud very well, can't find every flaw in a paper's methodology or conclusions, and takes up an enormous amount of scientists' time. But as we have seen throughout this book, peer review has been shaped just as much by social and political concerns as it has by epistemological ones. The first journals to adopt refereeing systems did so in response to concerns about editorial workloads or the wish to protect the reputation and collegiality of a learned society. During the Cold War, the idea that peer review is a prerequisite for scientific legitimacy did not arise because scientists noticed a marked difference in quality between refereed and unrefereed journals or grant programs, but because elevating peer review enabled scientists to defend their role in advising the government about scientific funding.

So what is peer review for in 2025? I don't think there's a clear single answer to that question—nor do I think there *should* be. As we saw in the second half of this book, "peer review" used to specifically mean grant review at the NIH, but it has since morphed into an umbrella term applied to review processes at a huge number of institutions. A shared term does not mean that every institution needs to use the same kind of peer review process or have the same goals for employing it. Some organizations might want their peer review to approve innovative and potentially disruptive work; others might emphasize work that shows potential for the greatest social benefit; still others might seek to approve everything that seems to be sound science, regardless of whether it is exciting or novel.[34] All of those are legitimate purposes for peer review. Institutions will, however, want to make sure that their expectations of peer review are realistic. An institution that adopts peer review in the hopes that it will root out fraud, for example, is setting itself up for failure. But there are other important goals a peer review process can serve well, such as making sure scientists will be consulted about funding decisions for scientific research and generating trust between the scientific community and other stakeholders (more on that below).

The question of accountability is also critical, and it's one that I think scientists sometimes forget to consider when they talk about changes they'd like to see. Many proposed reforms to peer review systems seem to assume that the only people with a stake in peer review—the only people to whom the institutions using it are accountable—are scientists. But that isn't the case. The idea that a paper or grant is "peer reviewed" is a major pillar of trust between the scientific community and stakeholders outside of science. An incident from July 2020 has long stuck in my mind. As the world scrambled to address the COVID-19 pandemic, the US Congress held a hearing on "The Urgent Need for a National Plan to Contain the Coronavirus." Anthony Fauci, the director of the National Institute of Allergy and Infectious Diseases (an NIH member institution), was called to testify about his efforts leading the pandemic response. During the hearings, the conservative Missouri congressman Blaine Luetkemeyer asked why Fauci was not endorsing the use of the lupus treatment drug hydroxychloroquine to treat COVID-19. He cited an "extensive study on hydroxychloroquine" that claimed the drug reduced the COVID-19 death rate. "The study was highly analyzed and peer-reviewed, unlike many other studies," Congressman Luetkemeyer informed Dr. Fauci.

Fauci replied that he was aware of the paper but called it a "flawed study." "I think anyone who examines it carefully sees that it is not a randomized, placebo-controlled trial," he explained.

Luetkemeyer interrupted, simply, "It said it was peer reviewed."[35]

Luetkemeyer's comments reflect the ideas about peer review that scientists themselves pushed in the 1970s: that review by scientific peers is the way to determine whether a piece of research is valuable. Members of the non-scientist public—including legislators and policymakers across the political spectrum—now see peer review as crucial to scientific legitimacy because they feel they can bestow greater trust on claims that experts have reviewed and critiqued before publication or funding. Reforms to peer review will need to take the question of public accountability seriously. Dramatically changing or discarding the practice of peer review would raise questions about which findings members of the public should now consider scientifically legitimate. If scientists wish to broaden the range of publications that might be considered "actual science," we will also have to consider how members of the public—including science journalists, a major conduit of information about science—are going to find their way through the resulting maze of preprints, published articles, website interviews, and social media posts. We should not assume that this will be an easy task for them to accomplish. Carole Lee has observed that dissemination of non-peer-reviewed papers via preprint servers increases what she calls the "certification-amplification gap," a discrepancy between which papers have withstood critique and scrutiny by other scientists and which scientific claims reach the largest audience.[36] Along similar lines, David Shatz has pointed to "inundation" as a potential consequence of eschewing prepublication peer review, a consequence that will make it difficult for all readers to find their way through the scientific literature.[37]

Some advocates of abolishing peer review argue that post-publication review via internet comments could fulfill the same certification functions as prepublication review, arguing that review by scientists after publication is not significantly different from review before publication.[38] Changing the stage at which refereeing occurs is a more profound shift than it might seem, however, because it involves a fundamental reimagining of what "peer reviewed research" is and where it can be found. That's not a change that can be made with a snap of the fingers and no follow-up questions. If I told my colleagues at the University of Maryland that this book was "peer reviewed," and they later learned that what I meant was that I had released a draft online and two other historians had posted comments about it, I doubt my department would praise me for embracing a cutting-edge definition of "peer review." They would feel misled, and rightly so. If scholars redefine what counts as peer review, they will have to explain and account for that change not just to other researchers, but to funders and members of the public as well.

Getting rid of peer review at funding bodies presents an additional set of challenges.[39] A preprint server can hold a theoretically infinite number of papers, but even the wealthiest foundation cannot afford to give money to every scientist who wants it. Choices between applications have to be made before money is given out. Some social scientists and philosophers have suggested that grants could operate via a lottery system, introducing an element of randomness to counteract biases against minorities, applicants from less prestigious institutions, and disruptive ideas. But once again, the accountability question throws a wrench into the works. At government funding bodies, legislators, bureaucrats, and members of the public alike will want to know how to ensure the money is being spent well, and that may be a hard question to answer if proposals are being funded based on chance. Private foundations might be able to experiment with lotteries with more success—in fact, some organizations like the British Academy have begun experimenting with grant programs that incorporate lotteries, often after an initial review to weed out obviously unsuitable proposals—but those foundations still have to convince their applicants, donors, and board members of the benefits of a system in which an outstanding proposal might lose out to a merely acceptable one because of a random number draw.[40]

As I have shown in this book, the idea that science must be peer reviewed to be legitimate is much younger than most observers assume. "Young," however, does not mean "easy to change or discard." Peer review is now so deeply embedded in both research institutions and in the public imagination that major changes to either journal or grant peer review would not be a simple discarding of a vestigial organ that no longer serves its purpose. Such changes would involve a complete overhaul of academic institutions and major shifts in the relationship between science and society at large. We're talking about a multi-system organ transplant, not an appendectomy. Which is not to say that large-scale change is impossible—but reformers should not underestimate the work required, especially if they do not want to break important things along the way.

PRESERVING THE BENEFITS OF PEER REVIEW

Which brings me back to the events of 2025. The US scientific and scholarly community is currently living through the Trump administration's rapid overhaul of both US peer review systems and the entire US research infrastructure. Trump administration officials, I think, would emphatically agree that modern peer review favors orthodox conclusions and established researchers. Their response is to slash funding to universities, shun top medical journals, found their own publications with small groups of approved reviewers, staff funding bodies with political appointees, and replace the members of major scientific advisory panels with researchers from outside the scientific mainstream.[41] I don't think any of that is what most reform-minded scientists envisioned when they complained about peer review reinforcing conventional ideas or letting flawed results into print, however. The past few months here in the US highlight that there are many possible paths for changes to peer review, and some of them might lead to places where scientists don't want to go.

The current moment also demonstrates that refereeing is, and has always been, a historically contingent phenomenon. The idea that a finding has to be peer reviewed to be "real science" was not handed down from on high during the early days of the Scientific Revolution. The idea that peer review was a "lynchpin" of science arose only in the late twentieth-century United States, and it took hold in response to a very specific set of historical circumstances, namely, conflicts about whether scientists ought to weigh in on how the US government spent its money on scientific research. Refereeing has undergone major changes in the past and will likely do so again in the future.

I wouldn't dare predict what peer review in the US might look like after our current set of historical circumstances, but we do seem to be at a moment when change is being thrust upon the scientific community, and those have been the moments at which refereeing in the past underwent its largest shifts.[42] If I have one recommendation to make based on my historical work, it is this: I think it's time to move past the idea that peer review makes something "actual science." Even the most thoughtful reviewers will occasionally fail to spot flaws in a proposal or paper, and referees' ability to recognize subtle problems grows more and more strained as the time they are able to devote to peer review shrinks. Peer review is, in my view, a useful minimum standard for taking a finding seriously, but it is the beginning of a conversation rather than the end—and acknowledging that there are some functions peer review can't perform well will help point to the most productive avenues for reform.

As scientists and scholars try to figure out their next steps, it's worth considering both ways that peer review could be improved and the things about it that they value and would like to preserve. Building public trust and preserving a role for experts in publication and funding decisions are functions that should not be discarded lightly. There can be other benefits as well. One of the most persuasive defenses of peer review I've read in recent months came, in fact, from the referee reports on this book manuscript. In their comments on my conclusion, Reviewer #3 argued that what peer review can provide is "mutuality among scientists," a commitment to reading and engaging with each others' work. At its best, peer review provides a mechanism for researchers to evaluate new findings and give honest feedback on a project's weaknesses and strengths. It helps spread word of forthcoming papers and recent achievements, creates shared intellectual investment in projects, and reminds us that the goal of research is not only individual advancement—it is to contribute to knowledge and to our research communities. For all peer review's flaws and frustrations and challenges, there are things about it that are worth defending.

ARCHIVAL COLLECTIONS CITED
(WITH ABBREVIATIONS)

David Lazarus Papers, University of Illinois Special Collections, Urbana-Champaign, Illinois (David Lazarus Papers).

Jacob J. Bikerman Papers, Niels Bohr Library and Archives, College Park, MD (Jacob J. Bikerman Papers)

John A. Hartford Foundation Records, Rockefeller Archives Center, Sleepy Hollow, New York (Hartford Foundation Records)

John Tyndall Correspondence, Royal Institution, London, UK (Tyndall-RI)

National Academies of Science Archives, Washington, DC (NAS Archives)

Papers of the National Science Foundation Historian, National Archives and Records Administration, College Park, MD (NSF Historian)

Papers of the NIH Director, National Archives and Records Administration, College Park, MD (NIH Director)

Papers of Sir George Gabriel Stokes, Add. MS 7656, Cambridge University Library, Cambridge, UK (Stokes Papers)

Philip Abelson Papers, American Association for the Advancement of Science Archives, Washington, DC (Abelson Papers)

Records of Hugh C. Wolfe, Niels Bohr Library and Archives, College Park, MD (Records of Hugh C. Wolfe)

Rockefeller Foundation Records, Rockefeller Archives Center, Sleepy Hollow, New York (RFR)

Royal Society Archives, Royal Society of London, London, UK (Royal Society Archives)

Samuel A. Goudsmit Papers, Niels Bohr Library and Archives, College Park, MD (Goudsmit Papers)

University of Maryland College Park University Archives, College Park, Maryland (UMD Archives)

W. F. Meggers Papers, Niels Bohr Library and Archives, College Park, MD (Meggers Papers)

NOTES

INTRODUCTION

1 This journal is now called *Notes and Records: The Royal Society Journal of the History of Science*. For the potentially curious, the paper is Melinda Baldwin, "'Where Are Your Intelligent Mothers to Come From?': Marriage and Family in the Scientific Career of Dame Kathleen Lonsdale FRS (1903–1971)," *Notes and Records of the Royal Society* 63 (2009): 81–94.

2 Hope Kelly, "Research Guides: Evaluate My Sources: Scholarly/Peer-Reviewed Sources," accessed December 20, 2023, https://guides.library.vcu.edu/evaluate/peer-review.

3 David Davies, quoted in Melinda Baldwin, *Making "Nature": The History of a Scientific Journal* (University of Chicago Press, 2015), 180.

4 This feature is not as useful as some searchers might think. If a scholarly journal uses peer review today, search engines tend to tag everything ever published in that journal as "peer reviewed"—including content published before that journal used peer review, as well as content such as book reviews or news articles that does not undergo peer review.

5 Sebastian Anthony, "CERN's Higgs Boson Discovery Passes Peer Review, Becomes Actual Science," ExtremeTech, 10 September 2012, https://www.extremetech.com/extreme/135756-cerns-higgs-boson-discovery-passes-peer-review-becomes-actual-science.

6 Flaminio Squazzoni, Elise Brezis, and Ana Marušić, "Scientometrics of Peer Review," *Scientometrics* 113, no. 1 (2017): 501–502, https://doi.org/10.1007/s11192-017-2518-4.

7 E.g., Daryl Chubin and Edward Hackett, *Peerless Science: Peer Review and U.S. Science Policy* (State University of New York Press, 1990); Frank Ingelfinger, "Peer Review in Biomedical Publication," *American Journal of Medicine* 56 (1974): 686–692; Sheila Jasanoff, "Peer Review in the Regulatory Process," *Science, Technology, & Human Values* 10 (1985): 20–32; Sheila Jasanoff, *The Fifth Branch: Science Advisers as Policymakers* (Harvard University Press, 1990), chap. 4; Sheila Jasanoff, "Quality Control and Peer Review in Advisory Science," in *The Politics of Scientific Advice: Institutional Design for Quality Assurance*, ed. Justus Lentsch and Peter Weingart (Cambridge University Press, 2011), 19–35; Bruno Latour and Steve Woolgar, *Laboratory Life: The Construction of Scientific Facts* (Princeton University Press, 1986); David Pontille and Didier Torny, "From Manuscript Evaluation to Article Valuation: The Changing Technologies of Journal Peer Review," *Human Studies* 38 (2015): 57–79; Anne Weller, *Editorial Peer Review: Its Strengths and Weaknesses*, (Information Today, 2001).

8 See, e.g., Harini Calamur and Roohi Ghosh, "Adapting Peer Review for the Future: Digital Disruptions and Trust in Peer Review," *Learned Publishing* 37, no. 1 (2024): 49–54; Kathleen Fitzpatrick, *Planned Obsolescence: Publishing, Technology, and the Future of the Academy*, 1st ed. (NYU Press, 2011); Jamie Shaw, "Peer Review, Innovation, and Predicting the Future of Science: The Scope of Lotteries in Science Funding Policy," *Philosophy of Science* 90, no. 5 (2023): 1297–1306; Chunli Wei et al., "What Does Open Peer Review Bring to Scientific Articles? Evidence from PLoS Journals," *Scientometrics : An International Journal for All Quantitative Aspects of the Science of Science, Communication in Science and Science Policy* 128, no. 5 (2023): 2763–2776; Robert Frodeman and Adam Briggle, "The Dedisciplining of Peer Review," *Minerva* 50 (2012): 3–19; J. Britt Holbrook, "The Use of Societal Impacts Considerations in Grant Proposal Peer Review: A Comparison of Five Models," *Technology and Innovation* 12 (2010): 213–224; J. Britt Holbrook and Robert Frodeman, "Peer Review and the Ex Ante Assessment of Societal Impacts," *Research Evaluation* 20, no. 3 (2011): 239–246; Marianne Noel, "Opening Up of Editorial Activities at Chemistry Journals. What Does Editorship Mean and What Does It Involve?," *Frontiers in Research Metrics and Analytics* 7 (2022): 747846–747846, https://doi.org/10.3389/frma.2022.747846.

9 Michèle Lamont, *How Professors Think: Inside the Curious World of Academic Judgment* (Harvard University Press, 2009).

10 David Shatz, *Peer Review: A Critical Inquiry* (Rowman & Littlefield, 2004).

11 Carole Lee, "Commensuration Bias in Peer Review," *Philosophy of Science* 82 (2015): 1272–1283; Carole Lee, Cassidy Sugimoto, Guo Zhang, and Blaise Cronin, "Bias in Peer Review," *Journal of the American Society for Information Science and Technology* 64, no. 1 (2013): 2–17.

12 J. C. Burnham, "The Evolution of Editorial Peer Review," *JAMA* 263, no. 10 (1990): 1323–1329, at p. 1323.

13 See, e.g., Imogen Clarke, "The Gatekeepers of Modern Physics: Periodicals and Peer Review in 1920s Britain," *Isis* 106 (2015): 70–93; Alex Csiszar, "Peer Review: Troubled from the Start," *Nature* 532 (21 April 2016): 306–308; Imogen Clarke and

Jim Mussell, "Conservative Attitudes to Old-Established Organs: Oliver Lodge and *Philosophical Magazine*," *Notes and Records: The Royal Society Journal of the History of Science* 69, no. 3 (2015): 321–336; Roberto Lalli, "'Dirty Work', but Someone Has to Do It: Howard P. Robertson and the Refereeing Practices of *Physical Review* in the 1930s," *Notes and Records: The Royal Society Journal for the History of Science* 70 (2016): 151–174; Marianne Noel, "Back to Disciplines: Exploring the Stability of Publication Regimes in Chemistry: The Case of the Journal of the American Chemical Society (1879–2010)," *Humanities & Social Sciences Communications* 7, no. 1 (2020): 1–13, https://doi.org/10.1057/s41599-020-00543-6; Lewis Pyenson, "Physical Sense in Relativity: Max Planck Edits the Annalen der Physik, 1906–1918," in *The Young Einstein: The Advent of Relativity*, ed. Lewis Pyenson (Adam Hilger, 1985), 194–214.

14 Alex Csiszar, *The Scientific Journal: Authorship and the Politics of Knowledge in the Nineteenth Century* (The University of Chicago Press, 2018); Aileen Fyfe, Noah Moxham, Julie McDougall-Waters, and Camilla Mørk Røstvik, *A History of Scientific Journals: Publishing at the Royal Society, 1665–2015* (UCL Press, 2022).

15 Natalie Aviles, *An Ungovernable Foe: Science and Policy Innovation in the U.S. National Cancer Institute* (Columbia University Press, 2024); Michael Barany, "A Postwar Guide to Winning a Science Grant," *Physics Today Online* (20 March 2018): https://doi.org/10.1063/PT.6.4.20180320a; Michael J. Barany, "Rockefeller Bureaucracy and Circumknowing Science in the Mid-Twentieth Century," *International Journal for History, Culture and Modernity* 7, no. 1 (2 November 2019): 779–796, https://doi.org/10.18352/hcm.592; Noortje Jacobs and Pieter Huistra, "Funding Bodies and Late Modern Science," *International Journal for History, Culture and Modernity* 7, no. 1 (2 November 2019): 887–898, https://doi.org/10.18352/hcm.584; Robert Kohler, *Partners in Science: Foundations and Natural Scientists, 1900–1945* (University of Chicago Press, 1991), 68–70; R. M. MacLeod, "The Royal Society and the Government Grant: Notes on the Administration of Scientific Research, 1849–1914," *The Historical Journal* 14 (1971): 323–358; Richard Mandel, *A Half Century of Peer Review, 1946–1996* (National Institutes of Health, 1996); George T. Mazuzan, "'Good Science Gets Funded . . .': The Historical Evolution of Grant Making at the National Science Foundation," *Science Communication* 14 (September 1992): 63–90; Robin Wolfe Scheffler, *A Contagious Cause: The American Hunt for Cancer Viruses and the Rise of Molecular Medicine* (The University of Chicago Press, 2019); Laura Stark, "Funding Panels as Declarative Bodies: Meritocracy and Paradoxes of Decision-Making in Modern Science," *International Journal for History, Culture and Modernity* 7, no. 1 (2 November 2019): 868–886, https://doi.org/10.18352/hcm.593.

16 Rafael Muñoz-Tamayo et al., "Seven Steps to Enhance Open Science Practices in Animal Science," *PNAS Nexus* 1, no. 3 (11 July 2022): pgac106, https://doi.org/10.1093/pnasnexus/pgac106.

17 Paul Ralph, "Practical Suggestions for Improving Scholarly Peer Review Quality and Reducing Cycle Times," *Communications of the Association for Information Systems* 38, no. 1 (2016): 274–283, https://doi.org/10.17705/1CAIS.03813.

18 Holly Else, "'Sexist' Peer Review Causes Storm Online." Times Higher Education (THE), April 30, 2015, https://www.timeshighereducation.com/news/sexist-peer-review-causes-storm-online/2020001.article.

19 Nyssa J. Silbiger and Amber D. Stubler, "Unprofessional Peer Reviews Disproportionately Harm Underrepresented Groups in STEM," *PeerJ* 7 (12 December 2019): e8247, https://doi.org/10.7717/peerj.8247.

20 See, e.g., "Reviewer 2 (@GrumpyReviewer2) / X," X (formerly Twitter), August 23, 2020, https://twitter.com/GrumpyReviewer2; "Reviewer 2 Must Be Stopped! | Facebook," accessed 29 December 2023, https://www.facebook.com/groups/reviewer2/.

21 E.g., Elizabeth Dzeng, "How Academia and Publishing Are Destroying Scientific Innovation: A Conversation with Sydney Brenner," *Kings Review*, 14 February 2014, https://www.kingsreview.co.uk/interviews/how-academia-and-publishing-are-destroying-scientific-innovation-a-conversation-with-sydney-brenner; Adam Mastroianni, "The Rise and Fall of Peer Review: Why the Greatest Scientific Experiment in History Failed, and Why That's a Great Thing," *Experimental History*, 13 December 2022, https://www.experimental-history.com/p/the-rise-and-fall-of-peer-review; Adrian Mulligan, Louise Hall, and Ellen Raphael, "Peer Review in a Changing World: An International Study Measuring the Attitudes of Researchers," *Journal of the American Society for Information Science and Technology* 64, no. 1 (2013): 132–161, https://doi.org/10.1002/asi.22798; Richard Smith, "Peer Review: A Flawed Process at the Heart of Science and Journals," *Journal of the Royal Society of Medicine* 99, no. 4 (April 2006): 178–182. Some cases of scientific fraud have involved scientists writing their own referee reports; see Cat Ferguson, Adam Marcus, and Ivan Oransky, "Publishing: The Peer-Review Scam," *Nature* 515, no. 7528 (1 November 2014): 480–482, https://doi.org/10.1038/515480a.

22 E.g., Lutz Bornmann, Rüdiger Mutz, and Hans-Dieter Daniel, "Gender Differences in Grant Peer Review: A Meta-analysis," *Journal of Infometrics*, 1 (2007): 226–238; Donna Ginther, Walter Schaffer, Joshua Schnell, Beth Masimore, Faye Liu, Laurel Haak, and Raynard Kington, "Race, Ethnicity, and NIH Research Awards," *Science*, 333 (19 August 2011): 1015–1019; Christine Wennerås and Agnes Wold, "Nepotism and Sexism in Peer-Review," *Nature* 387 (22 May 1997): 341–343.

23 E.g., Paul Nightingale and Alister Scott, "Peer Review and the Relevance Gap: Ten Suggestions for Policy-Makers," *Science and Public Policy* 34 (2007): 543–553; Kyle Siler, Kirby Lee, and Lisa Bero, "Measuring the Effectiveness of Scientific Gatekeeping," *Proceedings of the National Academy of Sciences*, 112, no. 2 (2015): 360–365.

CHAPTER 1

1 Harriet Zuckerman and Robert K. Merton, "Patterns of Evaluation in Science: Institutionalisation, Structure and Functions of the Referee System," *Minerva: A Review of Science, Learning and Policy* 9, no. 1 (1971): 66–100.

2 See Noah Moxham and Aileen Fyfe, "The Royal Society and the Prehistory of Peer Review, 1665–1965," *The Historical Journal* 61, no. 4 (2018): 863–889, particularly page 867.

3 See Aileen Fyfe, Noah Moxham, Julie McDougall-Waters, and Camilla Mørk Røstvik, *A History of Scientific Journals: Publishing at the Royal Society, 1665–2015* (UCL Press, 2022), chaps. 1–2.

4 On the *Philosophical Transactions*, see Dwight Atkinson, *Scientific Discourse in Sociohistorical Context: The Philosophical Transactions of the Royal Society of London, 1675–1975* (L. Erlbaum Associates, 1999); Adrian Johns, *The Nature of the Book: Print and Knowledge in the Making* (University of Chicago Press, 1998), chap. 3; Adrian Johns, "Miscellaneous Methods: Authors, Societies and Journals in Early Modern England," *British Journal for the History of Science* 33, no. 2 (2000): 159–186, https://doi.org/10.1017/S0007087499003933; Fyfe et al., *History of Scientific Journals*. On the *Journal des Sçavans*, see Pierre Costabel, "'L'à-peu-près n'est pas leur fait': Journal Des Sçavans 12 Mai 1698," in *L'à-peu-près: Aspects anciens et modernes de l'approximation*, ed. Louis Frey et al. (École des Hautes Études en Sciences Sociales, 1988), 79–85; Betty Trebelle Morgan, *Histoire du Journal des sçavans depuis 1665 jusqu'en 1701* (Les Presses Universitaires de France, 1929). Other scientific periodicals from this era include the *Acta Eruditorum*; see A. H. Laeven, *The "Acta Eruditorum" under the Editorship of Otto Mencke (1644–1707): The History of an International Learned Journal Between 1682 and 1707* (APA—Holland University Press, 1990).

5 Alex Csiszar, *The Scientific Journal: Authorship and the Politics of Knowledge in the Nineteenth Century* (The University of Chicago Press, 2018), 5.

6 "An Account of the Invention of Grinding Optick and Burning-Glasses, of a Figure Not-Spherical, Produced Before the Royal Society," *Philosophical Transactions* 3 (1668): 631–632.

7 Jacob Gregory, "An Account of a Controversy Betwixt Stephano de Angelis, Professor of the Mathematicks in Padua, and Joh. Baptista Riccioli Jesuite as It Was Communicated Out of Their Lately Printed Books, by That Learned Mathematician Mr. Jacob Gregory, a Fellow of the R. Society," *Philosophical Transactions* 3 (1668): 693–698.

8 "Queries Concerning Vegetation, Especially the Motion of the Juyces of Vegetables, Communicated by Some Curious Persons," *Philosophical Transactions* 3 (1668): 797–801.

9 For experimental or observational papers, see, e.g., Samuel Colcpresse, "An Account of Some Observations Made by Mr. Samuel Colepresse at and Nigh Plimouth, An. 1667. by Way of Answer to Some of the Quaeries Concerning Tides, Propos'd Numb. 17. and 18," *Philosophical Transactions* 3 (1668): 632–634; Viscount Brouncker, "The Squaring of the Hyperbola, by an Infinite Series of Rational Numbers, Together with Its Demonstration, by That Eminent Mathematician, the Right Honourable the Lord Viscount Brouncker," *Philosophical Transactions* 3 (1668): 645–649; Monsieur Justel, "A New Discovery Touching Vision," *Philosophical Transactions* 3 (1668): 668–671.

10 "Histoire de S. Louis Roy de France Ecrite par le Sire de Joinville, & enrichie de novelles Observations & de plusiers anciennes Pieces par Charles du Fresne Sieur du Cange," *Journal des Scavans* (23 January 1668): 3–6.

11 M. Fabricius, "Extrait du journal d'Angleterre, Contenant quelques Nouvelles experiences de l'Infusion des Medicamens ans les veines," *Journal des Scavans* (23 January 1668): 8–9.

12 "Extrait d'une Lettre de M. Hugens a l'auteur du Journal, touchant la Réponse que M. Gregory à faite à l'examen du Livre intitule *Vera Circuli & Hyperboles Quadratura*, dont on a parlé dans le V. Journal de cette année," *Journal des Scavans* (11 November 1668): 77–80.

13 "Lettre de M. Pecquet sur la Nouvelle Découverte touchant la Veuë," *Journal des Scavans* (17 September 1668): 61–63.

14 The closest modern analogues might be *Nature, Science,* and other journals that publish both experimental papers and correspondence, book reviews, and news articles.

15 Mario Biagoli, for instance, has pointed to records indicating that the Royal Society was tasked with reviewing books and journals prior to publication. See Mario Biagioli, "From Book Censorship to Academic Peer Review," *Emergences: Journal for the Study of Media & Composite Cultures* 12 (2002): 11–45. However, more recent historical work has shown that in practice, such review did not actually take place in any systematic or meaningful way. See Moxham and Fyfe, "Royal Society," 864–866.

16 Moxham and Fyfe, "Royal Society," 863.

17 Moxham and Fyfe, "Royal Society," 869.

18 Moxham and Fyfe, "Royal Society," 872.

19 See Fyfe et al., *History of Scientific Journals*, 200–201.

20 Csiszar, *Scientific Journal*, 29.

21 James E. McClellan III, *Specialist Control: The Publications Committee of the Académie Royale Des Sciences (Paris), 1700–1793* (American Philosophical Society, 2003), 12.

22 See McClellan, *Specialist Control*, chap. 2.

23 McClellan, *Specialist Control*, 32.

24 McClellan, *Specialist Control*, 13.

25 See McClellan, *Specialist Control*, 32–33.

26 For further discussion of the Académie and its publications, see Csiszar, *Scientific Journal*, chaps. 1 and 2. The gendered language here is deliberate, as the Académie was restricted to men at this time.

27 Csiszar, *Scientific Journal*, 4.

28 "Man of science" was the term preferred by English-speaking scientific practitioners in the 1820s and 1830s. See Sydney Ross, "Scientist: The Story of a Word," *Annals of Science* 18, no. 2 (1962): 65–85.

29 See Csiszar, *Scientific Journal*, 135–136.

30 See Csiszar, *Scientific Journal*, 134–135.

31 Augustus Granville, quoted in Fyfe et al., *History of Scientific Journals*, 274. Granville's pamphlet was part of a debate incited by Charles Babbage, *Reflections on the Decline of Science in England and on Some of Its Causes* (B. Fellowes, 1830).

32 Fyfe et al., *History of Scientific Journals*, 274–276.

33 On the founding of *Proceedings*, see Fyfe et al., *History of Scientific Journals*, 263–273; Csiszar, *Scientific Journal*, chap. 3; Alex Csiszar, "Proceedings and the Public: How a Commercial Genre Transformed Science," in *Science Periodicals in Nineteenth-Century Britain: Constructing scientific communities*, ed. Gowan Dawson et al. (University of Chicago Press, 2020), 103–134.

34 Council of the Royal Society, quoted in Fyfe et al., *History of Scientific Journals*, 264.

35 On Whewell, see Laura J. Snyder, *The Philosophical Breakfast Club: Four Remarkable Friends Who Transformed Science and Changed the World* (Broadway Books, 2011).

36 On Whewell's idea, see Alex Csiszar, "Peer Review: Troubled from the Start," *Nature* 532 (21 April 2016): 306–308.

37 John Lubbock, quoted in Csiszar, *Scientific Journal*, 144.

38 George Airy, quoted in Csiszar, *Scientific Journal*, 144.

39 William Whewell and John William Lubbock, "Report: On an Inequality of Long Period in the Motions of the Earth and Venus," *Proceedings of the Royal Society of London* 3 (1837): 108–113.

40 Fyfe et al., *History of Scientific Journals*, 278.

41 For a detailed breakdown, see Fyfe et al., *History of Scientific Journals*, 279–281.

42 Csiszar, *Scientific Journal*, 150.

43 Fyfe et al., *History of Scientific Journals*, 282.

44 The requirements were further narrowed in an 1847 reform. See Marie Boas Hall, *All Scientists Now: The Royal Society in the Nineteenth Century* (Cambridge University Press, 1984), chap. 3.

45 On Royal Society reform in the early nineteenth century, see Boas Hall, *All Scientists Now*, 4–5; Maurice Crosland, "Explicit Qualifications as a Criterion for Membership of the Royal Society: A Historical Review," *Notes and Records of the Royal Society of London* 37 (1983): 167–187; Csiszar, *Scientific Journal*, chap. 3; Fyfe et al., *History of Scientific Journals*, chap. 8.

46 See Boas Hall, *All Scientists Now*, 83–88.

47 On Stokes's service to the Royal Society, see Sloan Evans Despeaux, "Stokes and the Royal Society," in *George Gabriel Stokes: Life, Science and Faith*, ed. Mark McCartney, Andrew Whitaker, and Alastair Wood (Oxford University Press, 2019), 141–151.

48 Like most men of science of his generation, Stokes held multiple paying positions at once, and income was a constant source of stress. See Fyfe et al., *History of Scientific Journals*, 300–302.

49 George Gabriel Stokes, quoted in Fyfe et al., *History of Scientific Journals*, 297.

50 Norman McMillan, "Ireland and the Reform of the Politics and Government of British Science and Education," in *Prometheus's Fire: A History of Scientific and Technological Education in Ireland*, ed. Norman McMillan (Tyndall Publications, 2000), 510.

51 David Wilson, *Kelvin and Stokes: A Comparative Study in Victorian Physics* (CRC Press, 1987), 208.

52 Aileen Fyfe et al., "Managing the Growth of Peer Review at the Royal Society Journals, 1865–1965," *Science, Technology, & Human Values* 45, no. 3 (2020): 405–429, at p. 414.

53 Aileen Fyfe and Anna Gielas, "Introduction: Editorship and the Editing of Scientific Journals, 1750–1950," *Centaurus* 62, no. 1 (2020): 5–20, https://doi.org/10.1111/1600-0498.12290, at p. 2.

54 Fyfe et al., "Growth of Peer Review," 411.

55 Fyfe et al., "Growth of Peer Review," 413.

56 Fyfe et al., *History of Scientific Journals*, 307.

57 See Fyfe et al., *History of Scientific Journals*, 300.

58 Michael Foster, quoted in Fyfe et al., *History of Scientific Journals,* 303.

59 On the relationship between Tyndall and Stokes, see Melinda Baldwin, "Tyndall and Stokes: Correspondence, Referee Reports, and the Physical Sciences in Victorian Britain," in *The Age of Scientific Naturalism: Tyndall and His Contemporaries*, ed. Bernard Lightman and Michael Reidy (Pickering and Chatto, 2014), 171–186.

60 See, for example, Stokes's exchanges with physicists Norman Lockyer and James Clerk Maxwell, as well as a surviving letter to an unknown recipient. Baldwin, "Tyndall and Stokes," 180–181.

61 George Gabriel Stokes to John Tyndall, 7 May 1861, Tyndall-RI, S221; John Tyndall, "The Bakerian Lecture: On the Absorption and Radiation of Heat by Gases and Vapours, and on the Physical Connexion of Radiation, Absorption, and Conduction," *Philosophical Transactions of the Royal Society of London* 151 (1861): 1–36.

62 Tyndall, "Bakerian Lecture," 1–36.

63 Tyndall, "Bakerian Lecture," 27.

64 Tyndall, "Bakerian Lecture," 33.

65 George Gabriel Stokes to William Thomson, 17 May 1861, printed in *The Correspondence Between Sir George Gabriel Stokes and Sir William Thomson, Baron Kelvin of Largs*, ed. David B. Wilson (Cambridge University Press, 1990), 272.

66 Thomson mentions refereeing Tyndall's paper in his correspondence with Stokes. See William Thomson to George Gabriel Stokes, 17 April 1861, printed in *The Correspondence Between Sir George Gabriel Stokes and Sir William Thomson, Baron Kelvin of Largs*, ed. Wilson, 270–271.

67 G. G. Stokes, "Report on Prof[r] Tyndall's Paper on the Relation of Radiant Heat to Aqueous Vapour," Royal Society of London, Royal Society Archives, London, RR_5_275.

68 George Gabriel Stokes to John Tyndall, 28 May 1863, Tyndall-RI, S232.

69 George Gabriel Stokes to John Tyndall, 11 June 1874, Tyndall-RI, S260; John Tyndall, "On the Atmosphere as a Vehicle of Sound," *Philosophical Transactions of the Royal Society of London* 164 (1874): 183–244.

70 G. G. Stokes, "Report on a paper by Dr Tyndall entitled On the atmosphere as a vehicle of sound," Royal Society of London, Royal Society Archives, London, RR_7_350.

71 George Gabriel Stokes to John Tyndall, 11 June 1874, Tyndall-RI, S260.

72 See Tyndall, "On the Atmosphere as a Vehicle of Sound," pages 206 and 215 for the most important examples.

73 George Gabriel Stokes to John Tyndall, 16 May 1861, Tyndall-RI, S222; John Tyndall to George Gabriel Stokes, 19 May 1861, Tyndall-RI, T1421; George Gabriel Stokes to John Tyndall, 22 May 1861, Tyndall-RI, S223.

74 George Gabriel Stokes to John Tyndall, 11 June 1874, Tyndall-RI, S260.

75 George Gabriel Stokes to John Tyndall, 11 June 1874, Tyndall-RI, S260.

76 John Tyndall to George Gabriel Stokes, 17 June 1874, Tyndall-RI, T1423.

77 George Gabriel Stokes to John Tyndall, 21 June 1874, Tyndall-RI, S261.

78 John Tyndall to George Gabriel Stokes, no date, Stokes Correspondence, T600.

79 I call journals and magazines for researchers "specialist scientific periodicals" to distinguish them from popular science magazines aimed at readers who were not themselves scientific researchers; however, we should acknowledge that those categories often overlapped, especially early in the nineteenth century. A useful introduction to periodicals about science in Victorian Britain is the edited volume Gowan Dawson et al., *Science Periodicals in Nineteenth-Century Britain: Constructing Scientific Communities* (University of Chicago Press, 2020).

80 Gowan Dawson and Jonathan R. Topham, "Scientific, Medical, and Technical Periodicals in Nineteenth-Century Britain: New Formats for New Readers," in *Science Periodicals in Nineteenth-Century Britain: Constructing Scientific Communities*, ed. Gowan Dawson et al. (University of Chicago Press, 2020), 38.

81 See Csiszar, *Scientific Journal*, chap. 2; Iain P. Watts, "'We Want No Authors': William Nicholson and the Contested Role of the Scientific Journal in Britain, 1797–1813," *British Journal for the History of Science* 47, no. 3 (2014): 397–419.

82 Danielle Farrier, "Funding Scientific Research: The Royal Society of London and the Development of Research Grants, 1800–1914" (PhD Thesis, University of St. Andrews, 2025), chap. 3.

83 See Melinda Baldwin, *Making "Nature": The History of a Scientific Journal* (University of Chicago Press, 2015), chaps. 2–4.

84 See Melinda Baldwin, "The Business of Being an Editor: Norman Lockyer, Macmillan and Company, and the Editorship of Nature, 1869–1919," *Centaurus* 62, no. 1 (2020): 111–124.

85 See George Gabriel Stokes to Norman Lockyer, 22 November 1872, reprinted in *Memoir and Scientific Correspondence of the Late Sir George Gabriel Stokes*, ed. Joseph Larmor, 2 vols. (Cambridge University Press, 1907), vol. 1, pp. 398–400; Joseph Larmor to Norman Lockyer, 24 June 1905, NLP, MSS 110.

86 See Imogen Clarke, "The Gatekeepers of Modern Physics: Periodicals and Peer Review in 1920s Britain," *Isis* 106, no. 1 (2015): 70–93.

87 Oliver Lodge, qtd. in Clarke, "Gatekeepers," 76.

88 See Clarke, "Gatekeepers," 75–76, 89–90.

89 On the significance of *Nature*'s publication speed to its contributors, see Melinda Baldwin, "'Keeping in the Race': Physics, Publication Speed and National Publishing Strategies in *Nature*, 1895–1939," *British Journal for the History of Science* 47 (2014): 257–279.

90 On *Nature*'s status as a commercial journal, see Baldwin, "Business of Being an Editor."

91 See Csiszar, *Scientific Journal*, 151.

92 Quoted in Csiszar, *Scientific Journal*, 153.

93 Quoted in Csiszar, *Scientific Journal*, 156.

94 See Aileen Fyfe, "Editors, Referees, and Committees: Distributing Editorial Work at the Royal Society Journals in the Late 19th and twentieth Centuries," *Centaurus* 62, no. 1 (2020): 125–140, https://doi.org/10.1111/1600-0498.12266.

CHAPTER 2

1 The paper accepted without refereeing was the famous Einstein-Podolsky-Rosen (EPR) paper outlining a series of thought experiments that the authors argued showed quantum mechanics to be an incomplete description of reality. Albert Einstein, Boris Podolsky, and Nathan Rosen, "Can Quantum-Mechanical Description of Physical Reality Be Considered Complete?" *Physical Review* 47 (1935): 777–780.

2 Albert Einstein, quoted in Daniel Kennefick, "Einstein Versus the *Physical Review*," *Physics Today* 58, no. 9 (September 2005): 43.

3 Henry Rowlands, quoted in Paul Lucier, "The Origins of Pure and Applied Science in Gilded Age America," *Isis* 103, no. 3 (September 2012): 527–536, https://doi.org/10.1086/667976, at 530–531.

4 John Michels, "Salutatory," *Science* 1 (3 July 1880): 6. For more on *Science*'s early years, see Sally Gregory Kohlstedt, "*Science:* The Struggle for Survival, 1880 to 1894," *Science* 209 (4 July 1980): 33–42.

5 Lawrence Badash, *Radioactivity in America: Growth and Decay of a Science* (Johns Hopkins University Press, 1979); Melinda Baldwin, *Making Nature: The History of a Scientific Journal* (University of Chicago Press, 2015), chap. 4.

6 Arthur Lamb, "The Journal of the American Chemical Society," in *A History of the American Chemical Society: Seventy-Five Eventful Years*, ed. Mary E. Weeks and Charles A. Browne (American Chemical Society, 1952), 305–336, at 326. For more on the history of ACS publications, see Marianne Noel, "Back to Disciplines: Exploring the Stability of Publication Regimes in Chemistry: The Case of the Journal of the American Chemical Society (1879–2010)," *Humanities & Social Sciences Communications* 7, no. 1 (2020): 1–13, https://doi.org/10.1057/s41599-020-00543-6; P. J. Stang, "124 Years of Publishing Original and Primary Chemical Research: 135,149

Publications, 573,453 Pages, and a Century of Excellence," *Journal of the American Chemical Society* 125, no. 1 (1 January 2003): 1–8, https://doi.org/10.1021/ja021403x.

7 Lamb, "Journal of the American Chemical Society," 327.

8 Paul Hartman, *A Memoir on The Physical Review: A History of the First Hundred Years* (American Institute of Physics, 1994), 85.

9 See Hartman, *Memoir*, chap. 7.

10 Alfred O. C. Nier and John H. Van Vleck, "John Torrence Tate, July 28, 1889–May 27, 1950," *Biographical Memoirs of the National Academy of Sciences* 47 (1975): 479. Mitchell also became Tate's second wife in 1945, six years after the death of his first wife Lois Fossler Tate.

11 Nier and van Vleck, "John Torrence Tate," 476.

12 David M. Dennison, "The Rotation of Molecules," *Physical Review* 28, no. 2 (1926): 318–333.

13 Nier and van Vleck, "John Torrence Tate," 471–472.

14 Roberto Lalli, "'Dirty Work', but Someone Has to Do It: Howard P. Robertson and the Refereeing Practices of Physical Review in the 1930s," *Notes and Records: The Royal Society Journal for the History of Science* 70 (2016): 151–174, at p. 11.

15 Lalli, "'Dirty Work,'" 8.

16 Lalli, "'Dirty Work,'" 7.

17 Lalli, "'Dirty Work,'" 8.

18 Lalli, "'Dirty Work,'" 15.

19 Lalli, "'Dirty Work,'" 16.

20 See Daniel J. Kevles, *The Physicists: The History of a Scientific Community in Modern America* (Harvard University Press, 1995), chap. 8.

21 Johannes-Geert Hagmann, "Mobilizing US Physics in World War I," *Physics Today* 70, no. 8 (August 2017): 44–50.

22 Quoted in Robert E. Kohler, *Partners in Science: Foundations and Natural Scientists, 1900–1945* (University of Chicago Press, 1991), 8. For more on the impact of Carnegie funding for science, see Kevles, *The Physicists*, 81–84.

23 Kevles, *The Physicists*, 82.

24 See Kevles, *The Physicists*, chap. 16.

25 On the Science Advisory Board, see Carroll W. Pursell, "The Anatomy of a Failure: The Science Advisory Board, 1933–1935," *Proceedings of the American Philosophical Society* 109, no. 6 (1965): 342–351.

26 Albert L. Barrows to Charles A. Wolverton, 14 September 1935, in Records of the Science Advisory Board, Folder: "EX Bd: Science Advisory Board, General, 1935," NAS Archives.

27 "Tentative Budget for Rockefeller Appropriation of $50,000," no date, in Records of the Science Advisory Board, Folder: "EX Bd: Science Advisory Board, Executive Com: Meetings, 1933–1935," NAS Archives; Memorandum from Karl T. Compton,

24 September 1935, in Records of the Science Advisory Board, Folder: "EX Bd: Science Advisory Board, General, 1935," NAS Archives. See also the documents in folders "Finance, Funds: Grants, Science Advisory Board," 1933–1935, in Records of the Science Advisory Board, Folder: "EX Bd: Science Advisory Board, General, 1935," NAS Archives.

28 Judith Jacobsen, *The Greatest Good: A History of The John A. Hartford Foundation* (The Hartford Foundation, 1984), 10–11.

29 On the history of the Ford Foundation, see Emily Hauptmann, "The Ford Foundation and the Rise of Behavioralism in Political Science," *Journal of the History of the Behavioral Sciences* 48, no. 2 (2012): 154–173; Patti W. Hunter, "Gertrude Cox in Egypt: A Case Study in Science Patronage and International Statistics Education During the Cold War," *Science in Context* 22 (2009): 47–83; John Krige, "The Ford Foundation, Physics and the National Security State. A Study in the Transnational Circulation of Knowledge," in *American Foundations and the Coproduction of World Order in the twentieth Century*, ed. John Krige and Helke Rausch (Vandenhoeck and Ruprecht, 2012), 189–209; Mark Solovey, *Shaky Foundations: The Politics-Patronage-Social Science Nexus in Cold War America* (Rutgers University Press, 2013).

30 On the Carnegie Corporation's pre–World War II natural sciences grants programs, see Kohler, *Partners in Science*, chaps. 4 and 5.

31 See Kohler, *Partners in Science*, chap. 9; William H. Schneider, "The Origin of the Medical Research Grant in the United States: The Rockefeller Foundation and the NIH Extramural Funding Program," *Journal of the History of Medicine and Allied Sciences* 70, no. 2 (2015): 279–311.

32 See "National Research Council—Report of the Chairman of the Council to the Administrative Committee concerning the activities of the Council during the period October, November, and December, 1934," in Records of the Science Advisory Board, Folder: "EX Bd: Science Advisory Board, General, 1934," NAS Archives.

33 "The Natural Science Budget," n.d., RG 3: Administration, Program and Policy, Box 23, Folder 171, 13–23, RFR.

34 See Kohler, *Partners in Science*, 296–297.

35 Kohler, *Partners in Science*, 266.

36 Several historians have credited molecular biology's rise as an area of scientific research directly to the Rockefeller Foundation, arguing that without the foundation's funding, the field might never have become a vibrant research area. See, e.g., Michel Morange, *A History of Molecular Biology* (Harvard University Press, 1998). Others have been more skeptical of crediting the foundation with the establishment of molecular biology, however; see, e.g., Lily E. Kay, *The Molecular Vision of Life: Caltech, the Rockefeller Foundation, and the Rise of the New Biology* (Oxford University Press, 1993).

37 Kohler, *Partners in Science*, 68–70.

38 *The Rockefeller Foundation Annual Report, 1929* (The Rockefeller Foundation, 1929), 213.

39 Frank Blair Hanson to William C. Boyd, 23 February 1945, Projects Series 200 Box 176, Folder 1616, RFR.

40 See, e.g., H. H. Plough to Warren Weaver, 7 November 1939, Projects Series 200 Box 127, Folder 1563, RFR; W. A. Perlzweig and Hans Neurath, "Outline of a Research Program in Protein Chemistry," [June 1939], Projects Series 200 Box 138, Folder 1702, RFR.

41 Warren Weaver, "Proposed Institute for Genetics and Development at Columbia University," 19 October 1939, Projects Series 200 Box 132, Folder 1630, RFR.

42 "Columbia University—Genetics | EC 2/21/57," Projects Series 200 Box 132, Folder 1630, RFR.

43 See Kohler, *Partners in Science*, 284–285.

44 [Warren Weaver], "Committee Paper No. II," page 2, Record Group 3, Series 915 Box 2, Folder 9, RFR. We should note that the relationship between the trustees and officers was not entirely easy during this period; the trustees expected officers to be able to account for their decisions, and foundation officers remembered presenting to the trustees as an adversarial experience. See Warren Weaver oral history, Volume 2, p. 296, Record Group 13 Box 13, RFR.

45 Michael J. Barany, "Rockefeller Bureaucracy and Circumknowing Science in the Mid-Twentieth Century," *International Journal for History, Culture and Modernity* 7, no. 1 (2 November 2019): 779–796, https://doi.org/10.18352/hcm.592, at p. 780.

46 Quotes from Michael J. Barany, "A Postwar Guide to Winning a Science Grant," 20 March 2018, *Physics Today Online*, https://doi.org/10.1063/PT.6.4.20180320a.

47 See, e.g., "Tuesday, February 24, 1938: Dr. Ludwig Kast, Josiah Macy, Jr., Foundation," Projects Series 200 Box 130, Folder 1605, RFR.

48 Warren Weaver oral history, Volume 3, p. 526, Record Group 13, Box 14, RFR.

49 Frank B. Hanson, "Diary: 1938," 103. Rockefeller Foundation records, Officers' Diaries, Record Group 12, F-L. Available online at https://dimes.rockarch.org/objects/kxn8QYDqEuRfCUpPZe2Rnm/view.

50 Alan Gregg, "Diary: 1938," 31. Rockefeller Foundation records, Officers' Diaries, Record Group 12, F-L. Available online at https://dimes.rockarch.org/objects/3nnHHzmHPYVPMkzc7BsvT2/view.

51 Warren Weaver, "Diary: 1934," 59. Rockefeller Foundation records, Officers' Diaries, Record Group 12, S-Z. Available online at https://dimes.rockarch.org/objects/8LsGhnn7kNHTg8ZeS23eny/view.

52 "Excerpt from letter from A.C. Chibnall to Warren Weaver," 14 February 1939, Projects Series 200 Box 130, Folder 1606, RFR. The full copy of the letter was not preserved in that file, suggesting that some of Chibnall's other comments may have been even more personal.

53 Warren Weaver, "Diary: 1932," 79. Rockefeller Foundation records, Officers' Diaries, Record Group 12, S-Z. Available online at https://dimes.rockarch.org/objects/YqsxDMWKyBekZsFrV6hEZM/view. For the award letter, see Norma S. Thompson

to A. Lawrence Lowell, 14 January 1933, Projects Series 200 Box 139, Folder 1719, RFR.

54 Frank B. Hanson, "Diary: 1 January—30 April 1936," 99, Rockefeller Foundation records, Officers' Diaries, Record Group 12, F-L. Available online at https://dimes.rockarch.org/objects/GJEQtxTv4nWymvmrhgh5Hf/view.

55 Frank B. Hanson, "Diary: 1939," 114. Rockefeller Foundation records, Officers' Diaries, Record Group 12, F-L. Available online at https://dimes.rockarch.org/objects/LBsFrNdynyRZHahSsb5mUk/view.

56 Warren Weaver, "Diary: 1937," 44. Rockefeller Foundation records, Officers' Diaries, Record Group 12, S-Z. Available online at https://dimes.rockarch.org/objects/WinwawQ2h8GMv3AcKvfFSZ/view.

57 Frank B. Hanson with Warren Weaver, "Connecticut College for Women | Friday, November 1, 1935," Projects Series 200 Box 135, Folder 1665, RFR. Creighton and McClintock both went on to decorated careers; Creighton became the president of the Botanical Society of America, and McClintock won the Nobel Prize in 1983.

58 Warren Weaver, "Diary: 1937," 221. Rockefeller Foundation records, Officers' Diaries, Record Group 12, S-Z. Available online at https://dimes.rockarch.org/objects/WinwawQ2h8GMv3AcKvfFSZ/view.

59 Pnina G. Abir-Am, "The Rockefeller Foundation and the Post-WW2 Transnational Ecology of Science Policy: From Solitary Splendor in the Inter-War Era to a Me Too Agenda in the 1950s," *Centaurus: International Magazine of the History of Mathematics, Science, and Technology* 52 (2010): 323–337, at p. 325.

60 Frank B. Hanson, "Diary: 1937," 175. Rockefeller Foundation records, Officers' Diaries, Record Group 12, F-L. Available online at https://dimes.rockarch.org/objects/LBsFrNdynyRZHahSsb5mUk/view.

61 The Foundation had some programs that designated money specifically for the support of historically black colleges and universities (HBCUs); those programs usually fell under the umbrella of the foundation's General Education Board rather than Weaver's Division of Natural Sciences. Foundation support for HBCUs was significant by the standards of the day but thoroughly dwarfed by the donations to predominantly white institutions. For example, in 1919, the Rockefeller Foundation gave $125,000 to the Meharry medical school and $250,000 to the Howard University medical school, arguing that "the provision of adequate facilities for the proper training of Negro physicians is urgent." However, that same year, the foundation gave $500,000 to the Yale medical school, $1.46 million to the University of Chicago medical school, and $1.8 million to the Johns Hopkins medical school. Andy Lazris, "A Covert Script: The AMA's Fight Against Racial Inclusion in US Healthcare" (master's thesis, University of Maryland Department of History, 2025), 127–128.

62 Warren Weaver, quoted in Barany, "A Postwar Guide to Winning a Science Grant."

63 Warren Weaver oral history, Volume 3, p. 524, Record Group 13, Box 14, RFR.

64 Warren Weaver, "Diary: 1934," 79. Rockefeller Foundation records, Officers' Diaries, Record Group 12, S-Z. Available online at https://dimes.rockarch.org/objects/8LsGhnn7kNHTg8ZeS23eny/view.

65 Frank B. Hanson, "J. B. Sumner (Biochemistry)," 29 January 1941, Projects Series 200, Box 135, Folder 1673, RFR.

66 Warren Weaver, "Diary: 1937," 61. Rockefeller Foundation records, Officers' Diaries, Record Group 12, S-Z. Available online at https://dimes.rockarch.org/objects/WinwawQ2h8GMv3AcKvfFSZ/view.

67 Frank Hanson, "Diary: 8 January—30 April 1934," 9. Rockefeller Foundation records, Officers' Diaries, Record Group 12, F-L. Available online at https://dimes.rockarch.org/objects/7T6A7DtCKGcoG7uGCJ3DYT/view.

68 Frank Blair Hanson to Harold Urey, 9 January 1940, Projects Series 200 Box 129, Folder 1586, RFR.

69 Frank Hanson to Rene Dubos, 9 July 1943, Projects Series 200 Box 176, Folder 1616, RFR.

70 Rene Dubos to Frank Hanson, 14 July 1943, Projects Series 200 Box 176, Folder 1616, RFR.

71 In evaluating these letters, it is important to note that only applicants who were awarded grants have files in the Rockefeller Foundation archives. It is likely that the foundation received negative feedback about some applicants, but files about non-awardees were not preserved, and so the tone of the surviving letters is overwhelmingly positive.

72 See various letters in Projects Series 200, Box 135, Folder 1673, RFR.

73 See, e.g., Hugh Taylor to Frank Hanson, 12 January 1940, Projects Series 200 Box 129, Folder 1586, RFR; J.C.W. Frazer to Frank Hanson, 24 January 1940, Projects Series 200 Box 129, Folder 1586, RFR.

74 Howard Lewis to Frank Hanson, 2 February 1936, Projects Series 200, Box 154, Folder 1886, RFR.

75 F. C. Koch to Frank Hanson, 7 March 1936, Projects Series 200 Box 154, Folder 1886, RFR.

76 See assorted documents in "Harvard University—Endocrinology | 1935–1936," Project Series 200, Box 140, folder 1726, RFR.

77 On institutional accountability and epistemology in the sciences, see Michael Lynch, "Science, Truth, and Forensic Cultures: The Exceptional Legal Status of DNA Evidence," *Studies in History and Philosophy of Science. Part C, Studies in History and Philosophy of Biological and Biomedical Sciences* 44, no. 1 (2013): 60–70, https://doi.org/10.1016/j.shpsc.2012.09.008; Michael Lynch and Simon Cole, "Science and Technology Studies on Trial: Dilemmas of Expertise," *Social Studies of Science* 35, no. 2 (April 2005): 269–311, https://doi.org/10.1177/0306312705048715. I thank one of my anonymous reviewers for this suggestion.

CHAPTER 3

1 "Minutes of the Joint Meeting of AAAS Editorial Board and Publications Committee, 10 June 1955," Box 1, Abelson Papers.

2 Quoted in David Kaiser, "The Postwar Suburbanization of American Physics," *American Quarterly* 56 (2004): 851–888, at p. 852. Candidly, this quote is probably famous only among historians of science.

3 $1.89 billion in 1945 translates into $34.25 billion in 2025.

4 Quoted in Melinda Baldwin, "Physics: The Musical?" *Physics Today Online* (16 August 2018): https://doi.org/10.1063/PT.6.4.20180816a.

5 David Kaiser, "Shut up and Calculate!," *Nature* 505 (2014): 153–155.

6 David Kaiser, *How the Hippies Saved Physics: Science, Counterculture, and the Quantum Revival* (W. W. Norton, 2012), 16.

7 David Kaiser, "The Physics of Spin: Sputnik Politics and American Physicists in the 1950s," *Social Research* 73 (2006): 1225–1252.

8 W. McNeil Lowry, "Graduate Aid Program—Interview with Dean Ralph Sawyer, Graduate School, University of Michigan," Ford Foundation Records, Series 1: Behavioral Sciences Office Files, 1951–1966, Box 2, folder 9, Rockefeller Archives Center, Sleepy Hollow, NY.

9 David Kaiser, "Cold War Requisitions, Scientific Manpower, and the Production of American Physicists After World War II," *Historical Studies in the Physical and Biological Sciences* 33, no. 1 (2002): 131–159, at p. 136.

10 Kaiser, "Cold War Requisitions," 136.

11 See Emily Gibson, "NSF and Postwar US Science," *Physics Today* 73, no. 5 (May 2020): 40–46; Daniel J. Kevles, "The National Science Foundation and the Debate over Postwar Research Policy, 1942–1945: A Political Interpretation of *Science—The Endless Frontier*," *Isis* 68, no. 1 (1977): 5–26; Milton Lomask, *A Minor Miracle: An Informal History of the National Science Foundation* (U.S. Government Printing Office, 1976), 39–53.

12 Vannevar Bush, *Science, the Endless Frontier: A Report to the President* (National Science Foundation, 2020 [1945]), 21.

13 Bush, *Science, the Endless Frontier*, 37.

14 On the social sciences at the NSF, see Mark Solovey, *Social Science for What? Battles over Public Funding for the "Other Sciences" at the National Science Foundation* (MIT Press, 2020).

15 See Gibson, "NSF and Postwar US Science," 43.

16 Alan T. Waterman, "National Science Foundation: A Ten-Year Résumé," *Science* 131, no. 3410 (6 May 1960): 1342.

17 George T. Mazuzan, "'Good Science Gets Funded . . .': The Historical Evolution of Grant Making at the National Science Foundation," *Science Communication* 14 (September 1992): 63–90.

18 See Boxes 5, 6, and 11, Meggers Papers.

19 W. F. Meggers to [National Science Foundation], no date, Box 5, Folder 1959, Meggers Papers.

20 Meggers's letter has not survived, but the reply from Stanley gives an idea of its contents.

21 Robert W. Stanley to W. F. Meggers, 8 June 1959, Box 5, Folder 1959, Meggers Papers.

22 William F. Meggers to K. W. Meissner, 31 July 1955, Box 11, Folder 1955, Meggers Papers.

23 K. W. Meissner to William F. Meggers, 23 July 1955, Box 11, Folder 1955, Meggers Papers.

24 J. Howard McMillen to W. F. Meggers, 16 May 1955, Box 11, Folder 1955, Meggers Papers.

25 See, e.g., George R. Harrison to William F. Meggers, 23 January 1940, Box 9, Folder 1940, Meggers Papers.

26 William F. Meggers to [National Science Foundation], 23 May 1955, Box 11, Folder 1955, Meggers Papers.

27 Memorandum from Louis Levin, Office of Program Development and Analysis, on "Proposed Uniform Quality Rating System," 8 December 1965, Box 25, NSF Historian.

28 "Appropriations (Section 1)" and "Appropriations (Section 2)," *The NIH Almanac*, National Institutes of Health, https://www.nih.gov/about-nih/what-we-do/nih-almanac/appropriations-section-2.

29 On the influence of the Rockefeller Foundation on the NIH, see William H. Schneider, "The Origin of the Medical Research Grant in the United States: The Rockefeller Foundation and the NIH Extramural Funding Program," *Journal of the History of Medicine and Allied Sciences* 70, no. 2 (2015): 279–311.

30 On the establishment of the NIH and the political maneuvering that went into its creation, see Daniel M. Fox, "The Politics of the NIH Extramural Program, 1937–1950," *Journal of the History of Medicine and Allied Sciences* 42, no. 4 (October 1987): 447–466. See also Richard Mandel, *A Half Century of Peer Review, 1946–1996* (National Institutes of Health, 1996), 16.

31 See Schneider, 293–294.

32 Mandel, *A Half Century of Peer Review*, 46.

33 Mandel, *A Half Century of Peer Review*, 47.

34 David Kaiser, "Information Overload," unpublished draft from the forthcoming *American Physics and the Cold War Bubble*.

35 In 1964–1965, Goudsmit briefly split *Physical Review* into two journals, *Physical Review A* and *Physical Review B*, but went back to a single unified journal before settling on the four-journal solution.

36 Today the American Institute of Physics is more commonly shorthanded as "AIP," but in 1958 the organization referred to itself as "the Institute."

37 "Research on Publishing Problems in Physics," *Physics Today* 11 (September 1958): 32.

38 Franz J. Ingelfinger, "Peer Review in Biomedical Publication," *The American Journal of Medicine* 56 (1974): 686–692.

39 The records of *Science*'s editorial board include many documents comparing *Science* to these two journals. See, e.g., "Minutes of Meeting of Editorial Board of AAAS | 22 January 1956," Box 1, Abelson Papers; "Minutes | Meeting of the Science Editorial Board | 8 March 1974," Box 2, Abelson Papers.

40 On the Cattell editorship, see Michael M. Sokal, "*Science* and James McKeen Cattell, 1894 to 1945," *Science* 209 (4 July 1980): 43–52.

41 See John Walsh, "*Science* in Transition, 1946 to 1962," *Science* 209 (4 July 1980): 53.

42 Walsh, "*Science* in Transition," 52–53.

43 "Report of the chairman, Editorial Board, | of | February 11 1951," p. 3, Box 1, Abelson Papers.

44 Bentley Glass, "Report of the Acting Chairman of the Editorial Board, 22 June 1953," p. 9, Box 1, Abelson Papers.

45 Howard Meyerhoff, "Minutes of the meeting of the Editorial Board | May 28 1950," p. 3, Box 1, Abelson Papers.

46 Glass, "Report."

47 Glass, "Report."

48 Bentley Glass to Members of the Editorial Board of SCIENCE, 18 September 1953, p. 1, Box 1, Abelson Papers.

49 "Minutes | Editorial Board Meeting | 31 May 1953," Box 1, Abelson Papers.

50 "Minutes of the Meeting of the Editorial Board of Science and The Scientific Monthly, held in Washington on Sunday, May 30, 1954," pp. 3–4, Box 1, Abelson Papers.

51 See Bentley Glass, "Letter to be sent to Vice Presidents, Secretaries of sections, and Council Members of the AAAS," 22 June 1953, Abelson Papers, Box 1; "Minutes of the Meeting of the Editorial Board of Science and The Scientific Monthly, held in Washington on Sunday, May 30, 1954," pp. 3–4, Box 1, Abelson Papers.

52 "Minutes of the Meeting of the Editorial Board of Science and The Scientific Monthly, held in Washington on Sunday, May 30, 1954," p. 3, Box 1, Abelson Papers.

53 "Minutes | Editorial Board Meeting | 31 May 1953," p. 2, Box 1, Abelson Papers.

54 "Minutes of the Joint Meeting of AAAS Editorial Board and Publications Committee, 10 June 1955," p. 2, Box 1, Abelson Papers.

55 See "Editorial Board Meeting, 1 June 1956," Box 1, Abelson Papers. On the DuShane editorship, see Walsh, "*Science* in Transition," 52–57.

56 "Science—Report to the Board of Directors | American Association for the Advancement of Science | 2 and 3 March 1963," page 13, Box 9, Abelson Papers.

57 "Meeting, Editorial Board of Science | 26 May 1966," p. 2, Box 1, Abelson Papers.

58 "Report of the Meeting of the Editorial Board of Science | 28 March 1969," p. 2, Box 1, Abelson Papers.

59 "Meeting, Editorial Board of Science | 26 May 1966," p. 2, Box 1, Abelson Papers.

60 "Instructions to Reviewers—Solicited Articles," Box 2, folder "Editorial Board—1966," Abelson Papers.

61 "Instructions to Reviewers—Solicited Articles."

62 "Science—Report to the Board of Directors | American Association for the Advancement of Science | 26 and 27 March 1966," p. 1, Box 2, folder "Editorial Board—1966," Abelson Papers.

63 "Meeting, Editorial Board of Science | 16 February 1968," p. 3, Box 1, Abelson Papers.

64 "Science: Report to the Board of Directors | American Association for the Advancement of Science | 17–18 March 1973," p. 7, Box 2, folder "Editorial Board—1972," Abelson Papers.

65 "Minutes | Editorial Board Meeting | 31 May 1953."

66 Samuel Goudsmit to John Torrence Tate, 16 September 1949, in Series 11, Box 50, Folder 42: "Goudsmit referee reports for *Physical Review*, 1946–1950," Goudsmit Papers.

67 Although the brevity of Goudsmit's report makes it difficult to say for certain, the paper he refereed was likely H. A. Wilson, "The Spherical Shell Nuclear Model," *Physical Review* 77 (February 1950): 516–520.

68 "Memo to: AIP Publication Board | From: Hugh C. Wolfe," 2 November 1961, Box 1, Folder "AIP Publication Board 62/61/60/59," Records of Hugh C. Wolfe.

69 "Meeting of the Publication Board—December 14, 1959," p. 5, Box 1, Folder "AIP Publication Board 62/61/60/59," Records of Hugh C. Wolfe.

70 François Frenkiel to James Krumhansl, 1 December 1960, Box 1, Folder "AIP Publication Board 62/61/60/59," Records of Hugh C. Wolfe.

71 In 1962, for example, *Physical Review* regularly missed its targeted printing date by over month. See "Managing Editor's Report for 1962," p. 5, Box 1, Folder "AIP Publication Board 62/61/60/59," Records of Hugh C. Wolfe.

72 "Table A-1 | Required Manuscript-to-Print Intervals | January–June, July–December—1962," Box 3, Folder "Publishing Operations Report—1962," Records of Hugh C. Wolfe.

73 "Meeting of the Publication Board—December 14, 1959," p. 5, Box 1, Folder "AIP Publication Board 62/61/60/59," Records of Hugh C. Wolfe. The reasons for not wanting a checklist were not explained in the meeting minutes.

74 "Meeting of the Publication Board | April 25, 1962 | Washington, D.C.," p. 2, Box 1, Folder "AIP Publication Board 62/61/60/59," Records of Hugh C. Wolfe.

75 See "Managing Editor's Report for 1962," page 4, Box 1, Folder "AIP Publication Board 62/61/60/59," Records of Hugh C. Wolfe.

76 *Reviews of Modern Physics*, which published review articles rather than original research, also did not have a referee form.

77 "Journal of Mathematical Physics," Box 5, Folder "NSF Grant Renewal Proposals," Records of Hugh C. Wolfe. This folder appears to have been mislabeled.

78 "Journal of the Optical Society of America," Box 5, Folder "NSF Grant Renewal Proposals," Records of Hugh C. Wolfe.

79 "AIP Inter-Office Memorandum," Box 5, Folder "NSF Grant Renewal Proposals," Records of Hugh C. Wolfe. Lindsay's comments are handwritten at the bottom of the memo.

80 "Physical Review Letters | Report of Referee," Box 5, Folder "NSF Grant Renewal Proposals," Records of Hugh C. Wolfe. This folder appears to have been mislabeled.

81 "Report of Referee, Journal of Chemical Physics" and "The Astronomical Journal," Box 5, Folder "NSF Grant Renewal Proposals," Records of Hugh C. Wolfe. This folder appears to have been mislabeled.

82 "The Physical Review | Report of Referee," Box 5, Folder "NSF Grant Renewal Proposals," Records of Hugh C. Wolfe. This folder appears to have been mislabeled.

83 "Managing Editor's Report for 1962," p. 8, Box 1, Folder "AIP Publication Board 62/61/60/59," Records of Hugh C. Wolfe.

84 The outlier was *Physical Review Letters,* which rejected around 50 percent of articles; however, many of those rejected letters later found a home in *Physical Review* as full-length articles. See editorial reports in Box 1, Folder "AIP Publication Board 62/61/60/59," Records of Hugh C. Wolfe.

85 "Applied Optics Memorandum," Box 5, Folder "NSF Grant Renewal Proposals," Records of Hugh C. Wolfe. This folder appears to have been mislabeled.

86 Fyfe et al., *History of Scientific Journals*, 503–507.

87 See Melinda Baldwin, *Making "Nature": The History of a Scientific Journal* (University of Chicago Press, 2015), chap. 6.

88 Howard K. Zimmerman to Jacob J. Bikerman, 23 June 1962, Box 1, Folder 2: "Professional correspondence (1943–1958)," Jacob J. Bikerman Papers.

89 The Bikerman papers include several copies of referee reports that Bikerman wrote from 1945–1970.

90 [Jacob J. Bikerman] to Howard Zimmerman, 4 July 1962, Box 1, Folder 2: "Professional correspondence (1943–1958)," Jacob J. Bikerman Papers.

91 Michael Moravcsik, "Private and Public Communications in Physics," *Physics Today* 18 (March 1965): 23–26, https://doi.org/10.1063/1.3047261, at p. 25–26. This system is similar in some ways to the scheme William Whewell proposed in the 1830s, although Whewell had planned to publish only favorable reports. It is also similar to modern proposals for "open" peer review, in which all papers are published alongside their referee reports. See, e.g., Tony Ross-Hellauer, "What Is Open Peer Review? A Systematic Review," *F1000Research* 6 (31 August 2017): 588, https://doi

.org/10.12688/f1000research.11369.2; Natalie Shoham and Alexandra Pitman, "Open versus Blind Peer Review: Is Anonymity Better than Transparency?," *BJPsych Advances* 27, no. 4 (July 2021): 247–254, https://doi.org/10.1192/bja.2020.61; Jonathan P Tennant, "The State of the Art in Peer Review," *FEMS Microbiology Letters* 365, no. 19 (1 October 2018): fny204, https://doi.org/10.1093/femsle/fny204.

CHAPTER 4

1 On the early history of the sociology of science, see Gary Werskey, *The Visible College: The Collective Biography of British Scientific Socialists of the 1930's* (Holt, Rinehart, and Winston, 1979); Gerardo Ienna and Giulia Rispoli, "The 1931 London Congress: The Rise of British Marxism and the Interdependencies of Society, Nature and Technology," *HoST—Journal of History of Science and Technology* 15, no. 1 (1 June 2021): 107–130; A. K. Mayer, "Fatal Mutilations: Educationism and the British Background to the 1931 International Congress for the History of Science and Technology," *History of Science* 40 (2002): 445–472.

2 On Polanyi, see Mary Jo Nye, *Michael Polanyi and His Generation: Origins of the Social Construction of Science* (University of Chicago Press, 2011); William T. Scott and Martin X. Moleski, *Michael Polanyi: Scientist and Philosopher* (Oxford University Press, 2005).

3 See, e.g., Alan T. Waterman, "Review of *Social Influences and Scientists* by Warren O. Hagstrom," *Science* 151, no. 3706 (1966): 61–64; Margaret Gowing, "Scientists Only Human in Their Quest," *Nature* 254 (1975): 217–218; Dudley Shapere, "Review of *The Sociology of Science: Theoretical and Empirical Investigation*," *Physics Today* vol. 27 (August 1974): 52–53; Charles Weiner, "Review of *Historical Studies in the Physical Sciences, Vols. 1, 2*," *Physics Today* 25 (1971): 56–57; Edward MacKinnon, "Review of *Perception, Theory and Commitment: The New Philosophy of Science* by H. I. Brown," *Physics Today* 31 (1978): 73–74.

4 See, e.g., Robert K. Merton, "The Machine, the Worker, and the Engineer," *Science* 105 (1947): 79–84; Robert K. Merton, "Science, Population and Society," *The Scientific Monthly* 44 (1937): 165–171; Thomas S. Kuhn, "Historical Structure of Scientific Discovery," *Science* 136 (1962): 760–764; Michael Polanyi, "Scientific Outlook: Its Sickness and Cure," *Science* 125 (1957): 480–84; Derek de Solla Price, "The Scientific Foundations of Science Policy," *Nature* 206 (1965): 233–238; George Sarton, "An Institute for the History of Science and Civilization," *Science* 45 (1917): 284–286; I. Bernard Cohen, "The History of Science," *Science* 114, no. 2973 (1951): 3a.

5 On the Kuhnian framing of the adoption of continental drift, see John Stewart, *Drifting Continents and Colliding Paradigms: Perspectives on the Geosciences Revolution* (Indiana University Press, 1990).

6 A survey in the early 2000s found that roughly 75 percent of both early and mid-career scientists still agreed that scientists ought to conduct themselves in accordance with Merton's norms. Melissa S. Anderson et al., "Extending the Mertonian Norms: Scientists' Subscription to Norms of Research," *The Journal of Higher Education* 81, no. 3 (1 May 2010): 366–393, https://doi.org/10.1353/jhe.0.0095.

7 William Broad and Nicholas Wade, *Betrayers of the Truth: Fraud and Deceit in the Halls of Science* (Simon & Schuster, 1983), 18.

8 Michael Polanyi, *Science, Faith and Society* (Oxford University Press, 1946), 37.

9 J. D. Bernal, *The Social Function of Science* (MIT Press, 1967 [1939]), 296.

10 Gordon Tullock, *The Organization of Inquiry* (Duke University Press, 1966), 143.

11 Warren O. Hagstrom, *The Scientific Community* (Basic Books, 1965), 16.

12 Hagstrom, *Scientific Community*, 19.

13 Alvin M. Weinberg, "Criteria for Scientific Choice," *Minerva* 1 (1963): 159–171, at p. 162.

14 Proponents of preprint and draft circulation argued that preprints could be shared faster than journal articles, making it possible to advance scientific knowledge and detect flaws in scientific work more quickly. Skeptics said there would be little to no quality control over drafts and argued that journals could not possibly sustain their production costs if the model became sharing free preprints at no cost to colleagues. For commentary on preprints, see, e.g., Walter Loveland, "The Information Race," *Science* 148 (1965): 314; Alan Mather, "The Information Race," *Science* 148 (1965): 314; Simon Pasternack, "Is Journal Publication Obsolescent?" *Physics Today* 19 (May 1966): 38–43; Michael Moravcsik, "A Debate on Preprint Exchange—Pro: Physics Information Exchange," *Physics Today* 19 (1966): 62–69; Simon Pasternack, "A Debate on Preprint Exchange—Con: Criticism of the Proposed Physics Information Exchange," *Physics Today* 19 (1966): 63–69; William D. Garvey and Belver C. Griffith, "Scientific Communication as a Social System," *Science* 157 (1967): 1011–1016.

15 Derek J. De Solla Price, "Ethics of Scientific Publication," *Science* 144 (8 May 1964): 655–657, at p. 656.

16 J. M. Ziman, *Public Knowledge: An Essay Concerning the Social Dimension of Science* (Cambridge University Press, 1968), 103.

17 Ziman, *Public Knowledge*, 111.

18 Ziman, *Public Knowledge*, 116.

19 Ziman, *Public Knowledge*, 111.

20 Craig Calhoun, "On Merton's Legacy and Contemporary Sociology," in *Robert K. Merton: Sociology of Science and Sociology as Science*, ed. Craig Calhoun (Columbia University Press, 2010), 3–4.

21 Robert Merton, "Science and the Social Order," *Philosophy of Science* vol. 5 no. 3 (1938): 321–337; Robert Merton, "A Note on Science and Democracy," *Journal of Legal and Political Sociology* vol. 1 (1942): 115–126.

22 Alex Csiszar, "Gaming Metrics Before the Game: Citation and the Bureaucratic Virtuosos," in *Gaming the Metrics: Misconduct and Manipulation in Academic Research*, edited by Mario Biagioli and Alexandra Lippman (MIT Press, 2020), 31–42, at p. 31–32.

23 See Rena F. Subotnik, "Harriet Zuckerman: Eminent Sociologist of Science," *Journal for the Education of the Gifted* 19, no. 3 (1996): 344–358.

24 Csiszar, "Gaming Metrics Before the Game," 34.

25 Robert Merton, "Priorities in Scientific Discovery: A Chapter in the Sociology of Science," *American Sociological Review* vol. 22 no. 6 (1957): 635–659; Robert K. Merton, "The Matthew Effect in Science," *Science* vol. 159 (5 January 1968): 56–63.

26 Robert K. Merton, "Behavior Patterns of Scientists," *American Scientist* vol. 57 no. 1 (Spring 1969): 1–23, at p. 11.

27 Merton wrote, "It is now belatedly evident to me that I drew upon the interview and other materials of the Zuckerman study to such an extent that, clearly, the paper should have appeared under joint authorship. A sufficient sense of distributive and commutative justice requires one to recognize, however belatedly, that to write a scientific or scholarly paper is not necessarily sufficient grounds for designating oneself as its sole author." Robert K. Merton, "The Matthew Effect in Science: II: Cumulative Advantage and the Symbolism of Intellectual Property," *Isis* 79 (1988): 606–623, at p. 607.

28 Zuckerman and Merton would also begin a romantic relationship in 1968; they eventually married in 1993.

29 Harriet Zuckerman and Robert K. Merton, "Patterns of Evaluation in Science: Institutionalisation, Structure and Functions of the Referee System," *Minerva: A Review of Science, Learning and Policy* 9, no. 1 (1971): 66–100, at p. 67.

30 Zuckerman and Merton, "Patterns of Evaluation," 73.

31 Zuckerman and Merton, "Patterns of Evaluation," 75.

32 Zuckerman and Merton, "Patterns of Evaluation," 77.

33 Zuckerman and Merton, "Patterns of Evaluation," 78.

34 Zuckerman and Merton, "Patterns of Evaluation," 80.

35 Zuckerman and Merton, "Patterns of Evaluation," 81.

36 Zuckerman and Merton, "Patterns of Evaluation," 82.

37 See chart on Zuckerman and Merton, "Patterns of Evaluation," 91.

38 Zuckerman and Merton, "Patterns of Evaluation," 88.

39 Zuckerman and Merton, "Patterns of Evaluation," 95.

40 Zuckerman and Merton, "Patterns of Evaluation," 98.

41 Zuckerman and Merton, "Patterns of Evaluation," 99–100.

42 Harriet Zuckerman and Robert K. Merton, "Sociology of Refereeing," *Physics Today* 24, no. 7 (July 1971): 28–33.

43 Harold L. Davis, "Something We Can Feel Proud of," *Physics Today* 24, no. 7 (July 1971): 7.

44 Simon Pasternack, "Is Journal Publication Obsolescent?," *Physics Today* 19, no. 5 (1966): 38–43. Pasternack was the editor of *Physical Review* at the time, and his answer to the question posed in his title was "no."

45 Subotnik, "Harriet Zuckerman," 349.

46 For examples of papers that repeat this story, see, e.g., Von Bakanic, Clark McPhail, and Rita Simon, "The Manuscript Review and Decision-Making Process,"

American Sociological Review 52 (1987): 631–642; Daryl Chubin and Edward Hackett, *Peerless Science: Peer Review and U.S. Science Policy* (State University of New York Press, 1990); Robert Frodeman and Adam Briggle, "The Dedisciplining of Peer Review," *Minerva* 50 (2012): 3–19; Frank Ingelfinger, "Peer Review in Biomedical Publication," *The American Journal of Medicine* 56 (1974): 686–692; Stephen Lock, "Does Editorial Peer Review Work?" *Annals of Internal Medicine* 121 (1994): 60–61; Michael Mahoney, "Publication Prejudices: An Experimental Study in Confirmatory Bias in the Peer Review System," *Cognitive Research and Therapy* 1 (1977): 161–175.

47 Bernard K. Forscher, "Rules for Referees," *Science* 150 (15 October 1965): 321.

48 Forscher, "Rules for Referees," 319.

49 Forscher, "Rules for Referees," 320.

50 J. A. Krumhansl, George L. Trigg, and Gene L. Wells, "Authors Take Note," *Physical Review Letters* 36, no. 16 (1976): 917, https://doi.org/10.1103/PhysRevLett.36.917.

51 Pasternack, "Is Journal Publication Obsolescent?" 41.

52 Psychology Department Faculty Memo on "Recommendations for Promotion and Tenure Practices," n.d., page 15, in Folder P/T Actions 1976–77 (Folder 2 of 3), Vice President for Academic Affairs P34:A:55b Box No. 1, UMD Archives.

53 "College of Human Ecology Criteria for Faculty Merit Increases, Retention, and Promotion, 1975/76," page 2, in Folder P/T Actions 1976–77 (Folder 3 of 3), Vice President for Academic Affairs P34:A:55b Box No. 1, UMD Archives.

54 "Report of Committee on Promotions to Tenure Criteria," 1 December 1969, page 11, in Folder P/T Actions 1976–77 (Folder 1 of 3), Vice President for Academic Affairs P34:A:55b Box No. 1, UMD Archives.

55 E.g., Thomas O'Toole, "Nation Is Warned of Health Crisis," *Washington Post-Times Herald* (21 November 1967): A1; Spencer Rich, "Social Security Bill Approved in Senate: Senate Drops Family Welfare Reforms in Bill," *Washington Post-Times Herald* (6 October 1972): A1; Eric Morgenthaler, "Medical Checkup: Local Panels to Review Doctors' Work and Bills Are Urged by Critics; Public, Private Health Plans Being Abused by Physicians, Backers of Panel Assert; Case of the Costly Earache," *Wall Street Journal* (16 December 1970): 1, 20; Jerry E. Bishop, "The Wrong Way to Fight Cancer," *Wall Street Journal* (16 July 1971): 6; "AMA Delegates Deride Aetna Life for Letter Impugning Doctor Fees; Firm's Bid to Pay Cost of Legal Challenges to Bills Criticized; Consultation on Charges Urged," *Wall Street Journal* (23 June 1972): 6; "Medical Peer Review," *New York Times* (17 December 1972): E10; Nancy Hicks, "A.M.A. Will Back Peer Review Plan, But Seeks Changes," *New York Times* (27 July 1974): 1.

56 *Biomedical Science and Its Administration: A Study of the National Institutes of Health* (The White House, 1965).

57 Joseph Cooper, "Onward the Management of Science: The Wooldridge Report," *Science* 148 (11 June 1965): 1433–1439, at p. 1434.

58 See, e.g., Philip M. Boffey, "Health Crisis: LBJ Panel Calls for Reshaping American Medicine," *Science* 158 (1967): 1160–1162; Marti Mueller, "Oral Contraceptives:

Government-Supported Programs Are Questioned," *Science* 163 (1969): 553–555; Barbara J. Culliton, "Dual Publication: 'Ingelfinger Rule' Debated by Scientists and Press," *Science* 176 (1972): 1403–1405; "Medical Ethics Examined," *Nature* 242 (1973): 7–8.

59 E.g., "Academy Critic Shoots Wildly but Hits Sore Spot," *Nature* 229 (1971): 151–152; "Has Science Never Had It So Good?" *Nature* 233 (1971): 169–172.

60 Colin Norman, "Peer-Review Under Review," *Nature* 258 (1975): 382.

61 Samuel Levy, "A Perspective on Medicaid," *New England Journal of Medicine* 281 (1969): 297–301.

62 E.g., Robert Katz, "Peer Review Involving Nonmembers of Medical Societies," *Journal of the American Medical Association* 213 (1970): 1686; A. Michael Pardue, "Peer Review," *Journal of the American Medical Association* 225 (1973): 67; Calvin Kunin and Herman Efron, "Guidelines for Peer Review: Veterans Administration Ad Hoc Interdisciplinary Advisory Committee on Antimicrobial Drug Usage," *Journal of the American Medical Association* 237 (1977): 1001–1002.

63 Irvine H. Page, "A Different Peer Review," *Journal of the American Medical Association* 225 (1973): 1240–1241.

64 Lois DeBakey and Selma DeBakey, "Journal Editors and the Press: Cooperation Not Conflict," *Journal of the American Medical Association* 245 (1981): 2296–2297

65 See "Lord Zuckerman Defends His Position," *Nature* 240 (1972): 247.

66 Zuckerman, "Cancer Research," *Nature* 240 (1972): 497–498.

67 Franz J. Ingelfinger, "To Impart the Precepts and the Instruction," *The Lancet* 292 no. 7571 (5 October 1968): 766–767.

68 Franz J. Ingelfinger, "Medical Literature: The Campus Without Tumult," *Science* (1970): 831–837, at p. 836.

69 "There Shall Be Weeping and Gnashing of Teeth," *New England Journal of Medicine* 282 (1970): 393–394.

70 Thomas Jukes, "Peer Review," *Nature* 265 (1977): 203.

CHAPTER 5

1 On the War on Cancer, see Robin Wolfe Scheffler, *A Contagious Cause: The American Hunt for Cancer Viruses and the Rise of Molecular Medicine* (University of Chicago Press, 2019); Kenneth E. Studer and Daryl E. Chubin, *The Cancer Mission: Social Contexts of Biomedical Research* (Sage Publications, 1980); Keith Wailoo, *How Cancer Crossed the Color Line* (Oxford University Press, 2011); Guy B. Faguet, *The War on Cancer: An Anatomy of Failure, a Blueprint for the Future* (Springer, 2005); Richard A. Rettig, *Cancer Crusade: The Story of the National Cancer Act of 1971* (Princeton University Press, 1977).

2 Office of Management and Budget, "The National Institutes of Health and National Institute of Mental Health Peer Review System" (1973): I-7, NN3-443-05-006 Box 7, file Comm 2-5-L, NIH Director.

3 *Biomedical Science and Its Administration: A Study of the National Institutes of Health* (The White House, 1965), 195.

4 A. A. Kaufman, "NIH Study Report | NIH Public Advisory Groups and the Peer Review System," 27 September 1972, pp. 6–7, NN3-443-05-006 Box 7, file Comm 2-5-L, NIH Director.

5 "Peer Review: Commentary on an OMB Paper | Prepared by the Office of the Director, National Institutes of Health | Appendix I: The OMB Issue Paper and Transmittal Memorandum," dated 3 April 1973, p. I-3, NN3-443-05-006 Box 7, file Comm 2-5-L, NIH Director.

6 On the FDA's review procedures for approving new drugs, see Daniel Carpenter, *Reputation and Power: Organizational Image and Pharmaceutical Regulation at the FDA* (Princeton University Press, 2010), https://doi.org/10.1515/9781400835119.

7 All quotes are from "Peer Review: Commentary," 1–10.

8 "DRAFT: The Peer Review System in the Research Grant Area" (25 January 1971), p. 1, NN3-443-05-006 Box 7, file Comm 2-5-L, NIH Director.

9 *Biomedical Science*, 4. The operating budget, notably, did not include the grant money that the NIH managed, only the funds necessary to pay for NIH staff and expenses.

10 *Biomedical Science*, 4 and 21.

11 *Biomedical Science*, 201.

12 For more on these controversies and their fallout for the NIH, see Laura Stark, *Behind Closed Doors: IRBs and the Making of Ethical Research* (University of Chicago Press, 2011), 144–158.

13 *The Administration of Grants by the National Institutes of Health: Hearings Before a Subcommittee of the Committee on Government Operations* (U.S. Government Printing Office, 1962), 14. See also Daniel Greenberg, *The Politics of Pure Science* (University of Chicago Press, 1999 [1967]), 278–279.

14 *The Administration of Grants by the National Institutes of Health*, 51–87.

15 Richard Mandel, *A Half Century of Peer Review, 1946–1996* (National Institutes of Health, 1996), 96–99.

16 The NCA included funds to establish new cancer research centers and a centralized data bank on cancer research. It also created an eighteen-member National Cancer Advisory Board and a three-member President's Cancer Panel to monitor the progress of the "war on cancer." For the text of the original bill, see "National Cancer Act of 1971," *Cancer.gov*, accessed 27 October 2023, https://www.cancer.gov/about-nci/overview/history/national-cancer-act-1971. See also Richard Rettig, *Cancer Crusade: The Story of the National Cancer Act of 1971* (Princeton University Press, 1977).

17 Scheffler, *A Contagious Cause*, 163.

18 "National Institutes of Health: History of Congressional Appropriations, 1970–1979," https://officeofbudget.od.nih.gov/pdfs/FY08/FY08%20COMPLETED/appic3806%20-%20transposed%20%2070%20-%2079.pdf.

19 Barbara J. Culliton, "Training Grants: Tied Up in Congress with Ethics Bills," *Science* 182, no. 4109 (1973): 265–266.

20 Mandel, *A Half Century of Peer Review*, 144; Robert Gillette and Barbara Culliton, "Briefing," *Science* 179, no. 4070 (1973): 260–261.

21 Edward David, "Science, Politics and Policy in the Nixon Administration," in *Presidential Science Advisors: Perspectives and Reflections on Science, Policy and Politics*, ed. Roger Pielke and Roberta A. Klein (Springer Netherlands, 2010), 29–36.

22 John Walsh, "Federal Science: Filling the Blanks in Policy and Personnel," *Science* 179 (2 February 1973): 456–458.

23 Nicholas Wade, "Science Adviser's Exit: What Does It Mean for Science Policy?" *Science* 179 (2 February 1973): 458–459.

24 Office of Management and Budget, "The National Institutes of Health and National Institute of Mental Health Peer Review System" (1973), p. I-3, NN3-443-05-006 Box 7, file Comm 2-5-L, NIH Director.

25 Office of Management and Budget, "The National Institutes of Health and National Institute of Mental Health Peer Review System" (1973), p. I-5, NN3-443-05-006 Box 7, file Comm 2-5-L, NIH Director.

26 Office of Management and Budget, "The National Institutes of Health and National Institute of Mental Health Peer Review System" (1973), p. I-7, NN3-443-05-006 Box 7, file Comm 2-5-L, NIH Director.

27 Trofim Lysenko was a Soviet agronomist who rejected the science of genetics. Lysenko managed to attract the attention and support of Soviet leader Josef Stalin for his theories of inheritance. The Soviet government's support for Lysenko's theories led to a disastrous agricultural famine and to the suppression of genetics research in the USSR, including the arrest and death of the geneticist Nikolai Vavilov. See Nikolai Krementsov, "A 'Second Front' in Soviet Genetics: The International Dimension of the Lysenko Controversy, 1944–1947," *Journal of the History of Biology* 29 (1996): 229–250; Kirill O. Rossianov, "Editing Nature: Joseph Stalin and the 'New' Soviet Biology," *Isis* 84 (1993): 728–745; Valerii Soifer, Leo Gruliow, and Rebecca Gruliow, *Lysenko and the Tragedy of Soviet Science* (Rutgers University Press, 1994); Audra Wolfe, *Freedom's Laboratory: The Cold War Struggle for the Soul of Science* (Johns Hopkins University Press, 2019), chaps. 2 and 3.

28 On reactions to Lysenko in the US, see Michael D. Gordin, *The Pseudoscience Wars: Immanuel Velikovsky and the Birth of the Modern Fringe* (University of Chicago Press, 2012), chap. 2; Wolfe, *Freedom's Laboratory*, chap. 1.

29 See Wolfe, *Freedom's Laboratory*, chap. 4.

30 On Project Camelot, see Ellen Herman, "Project Camelot and the Career of Cold War Psychology," in *Universities and Empire: Money and Politics in the Social Sciences During the Cold War*, ed. Christopher Simpson (New Press, 1998), 97–133.

31 On Vietnam-era campus activism and science, see Stuart Leslie, *The Cold War and American Science: The Military-Industrial-Academic Complex at MIT and Stanford* (Columbia University Press, 1993); Matt Wisnioski, "Inside 'the System': Engineers,

Scientists, and the Boundaries of Social Protest in the Long 1960s," *History and Technology* 19 (2003): 313–333.

32 Leslie, *Cold War and American Science*, 233.

33 On Science for the People, see Donna Haraway, "The Transformation of the Left in Science: Radical Associations in Britain in the 30s and the USA in the 60s," *Soundings* 58 (1975): 441–462; Kelly Moore, "Organizing Integrity: American Science and the Creation of Public Interest Organizations, 1955–1975," *American Journal of Sociology* 101 (1996): 1592–1627; and Kelly Moore, *Disrupting Science: Social Movements, American Scientists, and the Politics of the Military, 1945–1975* (Princeton University Press, 2008). See also *Science*'s entertaining (though perhaps not entirely neutral) account of the 1970 Chicago meeting: Philip M. Boffey, "AAAS Convention: Radicals Harass the Establishment," *Science* 171 (8 January 1971): 47–49.

34 "Peer Review: Commentary," 22.

35 "Peer Review: Commentary," 18.

36 "Peer Review: Commentary," 11.

37 "Peer Review: Commentary," 23.

38 Barbara Culliton, "Peer Review: OMB May Dismantle NIH Study Sections," *Science* 180 (25 May 1973): 843–844.

39 Sydney Pulver to Caspar Weinberger (11 June 1973), NN3-443-05-006 Box 7, file Comm 2-5-L, NIH Director.

40 Jozef J. Zwislocki to Senator James L. Buckley (8 May 1973), NN3-443-05-006 Box 7, file Comm 2-5-L, NIH Director.

41 T. C. Hau to Congressman Paul G. Rogers (26 April 1973), NN3-443-05-006 Box 7, file Comm 2-5-L, NIH Director.

42 New York Academy of Sciences to the President (30 May 1973), NN3-443-05-006 Box 7, file Comm 2-5-L, NIH Director.

43 Charles Edwards, quoted in Barbara Culliton, "Peer Review: Edwards Denies System Will Be Undone," *Science* 180 (8 June 1973): 1035–1036.

44 Barbara Culliton, "NIH: Who Is Running the Show—Scientists or Politicians?" *Science* 183 (1 March 1974): 829–832.

45 Stuart Auerbach, "NIH Aide Resigns, Hits Policy," *Washington Post* (18 January 1974): A2.

46 "Trouble at NIH," *Washington Post* (19 January 1974): A14.

47 Charles Edwards, "On the Question of Autonomy for NIH," *Washington Post* (24 January 1974): A29.

48 Barbara Culliton, "Office of Management and Budget: Skeptical View of Scientific Advice," *Science* 183 (1 February 1974): 392–396.

49 Committee on Labor and Public Welfare Subcommittee on Health, *Quality of Health Care—Human Experimentation, 1973: Hearings Before the Subcommittee on*

Health of the Committee on Labor and Public Welfare (U.S. Govt. Print. Off., 1973), hereafter *Quality of Health Care.*

50 On the Tuskegee scandal, see Christopher Crenner, "The Tuskegee Syphilis Study and the Scientific Concept of Racial Nervous Resistance," *Journal of the History of Medicine and Allied Sciences* 67, no. 1 (2012): 244–280; Jonathon Erlen and Thomas G. Benedek, "The Scientific Environment of the Tuskegee Study of Syphilis, 1920–1960," *Perspectives in Biology and Medicine* 43, no. 1 (1999): 1–30; Susan M. Reverby, *Examining Tuskegee: The Infamous Syphilis Study and Its Legacy* (University of North Carolina Press, 2013).

51 *Quality of Health Care*, 1196.

52 *Quality of Health Care*, 1263.

53 *Quality of Health Care*, 19. It should be noted that several uses of "peer review" in these hearings referred to local review of physicians by other physicians, an older use of the term. See chapter 4 for a fuller history of the term.

54 See, e.g., "Senator Kennedy's Talk at Masur Auditorium," 23 May 1973, NN3-443-05-006 Box 7, file Comm 2-5-L, NIH Director. Attacking the Nixon administration's policy toward the NIH seems to have been a major motivation for Kennedy's role in the hearings. At several points during the hearings, Kennedy used his platform to criticize the Nixon administration's recent run-ins with the NIH, particularly the budget cuts. See, e.g., *Quality of Health Care*, 1255–1256.

55 There is, of course, a massive literature on the Watergate scandal. Good starting points include Carl Bernstein and Bob Woodward, *All the President's Men* (Simon & Schuster, 1974), the now-classic book by the *Washington Post* reporters who helped prove the cover-up; John Farrell, *Richard Nixon: The Life* (Vintage, 2017); Garrett Graff, *Watergate: A New History* (Simon & Schuster, 2023); and Stanley Kutler, *The Wars of Watergate: The Last Crisis of Richard Nixon* (Alfred A. Knopf, 1990).

CHAPTER 6

1 C. W. Sherwin, *First Interim Report on Project Hindsight* (Office of the Director of Defense Research and Engineering, 1966), 13.

2 Sherwin, *First Interim Report*, 11.

3 Sherwin, *First Interim Report*, 12.

4 On Project Hindsight, see Glen Ross Asner, "The Cold War and American Industrial Research," PhD diss., Carnegie Mellon University, 2006; Ronald Kline, "Construing 'Technology' as 'Applied Science': Public Rhetoric of Scientists and Engineers in the United States, 1880–1945," *Isis* 86 (1995): 194–221; Karl Kreilkamp, "Hindsight and the Real World of Science Policy," *Science Studies* 1 (1971): 43–66; Audra J. Wolfe, "The Military-Academic-Industrial Complex and the Path Not Taken," in *Origins of the National Security State and the Legacy of Harry S Truman*, ed. Mary Ann Heiss and Michale J. Hogan (Truman State University Press, 2015), 143–164.

5 The US "victory" in the space race was, of course, not quite so cut-and-dried. The Soviet Union had been first to launch a satellite, first to put a man into orbit, and first to put a woman into orbit. On the space race and Project Apollo, see Teasel Muir-Harmony, *Operation Moonglow: A Political History of Project Apollo* (Basic Books, 2020); Walter McDougall, *The Heavens and the Earth: A Political History of the Space Age* (Johns Hopkins University Press, 1997); Asif Siddiqi, *Challenge to Apollo: The Soviet Union and the Space Race, 1945–1974* (NASA History Office, 2000).

6 For the full data on US R&D expenditures, see "National Patterns of R&D Resources | NSF—National Science Foundation," accessed 2 January 2024, https://ncses.nsf.gov/data-collections/national-patterns/2021#data.

7 David Kaiser, *How the Hippies Saved Physics: Science, Counterculture, and the Quantum Revival* (W. W. Norton, 2012), 22–23.

8 Susanne D. Ellis, "Where Does He Come from? Where Does He Go?," *Physics Today* 22, no. 3 (1 March 1969): 53–57, https://doi.org/10.1063/1.3035461.

9 Kaiser, *How the Hippies Saved Physics*, 22–23. See also David Kaiser, "Booms, Busts, and the World of Ideas: Enrollment Pressures and the Challenge of Specialization," *Osiris* 27, no. 1 (January 2012): 276–302, https://doi.org/10.1086/667831.

10 Luther Carter, "President Ford: Main Street to Pennsylvania Avenue," *Science* 185 (1974): 764–766, at p. 764.

11 Carter, "President Ford," 765.

12 Barbara Culliton, "Biomedical Research Budget Drops Again," *Science* 187 (1975): 520–521.

13 See, e.g., The Editors to Dr. James Bonner, 9 November 1971, Abelson Box 90, Folder: Mayfield, S.E., Chromosomal RNA, Abelson Papers.

14 Conlan would go on to lose one of the most bitterly contended Senate primary races in US history. His opponent in the 1976 Arizona Republican senatorial primary, Sam Steiger, was a Jewish congressman. Conlan portrayed himself as a leader who would be the first in a wave of Christians elected to public office, and Conlan's campaign was accused of both subtle and overt antisemitism. Steiger won the primary, but Arizona Republicans were deeply divided by the viciousness of the campaign. Ultimately, despite Arizona's heavily Republican voter base, Steiger lost the Senate seat to the inexperienced Democratic candidate Dennis DeConcini. On Conlan, see John Turner, "Selling Jesus to Modern America: Campus Crusade for Christ, Evangelical Culture, and Conservative Politics," PhD diss., University of Notre Dame, 2006.

15 On MACOS, see Peter Dow, *Schoolhouse Politics: Lessons from the Sputnik Era* (Harvard University Press, 1991); R. W. Evans, *The Tragedy of American School Reform: How Curriculum Politics and Entrenched Dilemmas Have Diverted Us From Democracy* (Palgrave Macmillan, 2011), chap. 5; George T. Mazuzan, "'Good Science Gets Funded . . .': The Historical Evolution of Grant Making at the National Science Foundation," *Science Communication* 14 (September 1992): 63–90; Erika Milam, "Public Science of the Savage Mind: Contesting Cultural Anthropology in the Cold War Classroom," *Journal of the History of the Behavioral Sciences* 49 (2013): 306–330;

Mark Solovey, *Social Science for What? Battles over Public Funding for the "Other Sciences" at the National Science Foundation* (MIT Press, 2020), chap. 5.

16 On Bauman, see his memoir *The Gentleman from Maryland: The Conscience of a Gay Conservative* (Arbor House, 1986). Bauman's promising political career came to an abrupt end when he was arrested for soliciting a teenage male prostitute.

17 "U.S. Senate: Senator William Proxmire," accessed 13 February 2024, https://www.senate.gov/artandhistory/history/minute/Senator_William_Proxmire.htm.

18 William Proxmire, quoted in Elaine Hatfield and G. W. Walster, *A New Look at Love* (University Press of America, 1978). Hatfield was one of the researchers whose work was being criticized in Proxmire's press release.

19 William Proxmire, quoted in Constance Holden, "'Fleece' Winner Sues Proxmire," *Science* 192 (21 May 1976): 769.

20 For NSF correspondence with members of the public (often via their representatives) on controversial grants, see, e.g., Ralph Kazarian to Mrs. Joseph Long, 12 May 1975, NSF Historian 6-27-208; Elizabeth Hunt to Clifford Case, 27 May 1975, NSF Historian 6-27-237; Patricia Nicely to Robert Mills, 30 May 1975, NSF Historian 6-27-239; Patricia Nicely to Sam Nunn, 30 May 1975, NSF Historian 6-27-244; Jack Kratchman to Joe Evins, 12 June 1975, NSF Historian 6-27-266.

21 See, e.g., Patricia Nicely to Marvin Esch, 28 April 1975, NSF Historian 6-27-182; Jack Kratchman to Mrs. Merlin Lenox, 5 September 1975, NSF Historian 6-27-455.

22 See, e.g., Elizabeth Hunt to Ken Hechler, 12 November 1975, NSF Historian 6-27-564.

23 See, e.g., T. George Harris, "Introduction," in Elaine Walster and G. William Walster, *A New Look at Love* (Addison-Wesley, 1978), p. viii; Harris calls Proxmire's comments a "cheap shot" that cost the pair their grant money.

24 Holden, "'Fleece' Winner Sues Proxmire," 769.

25 See Solovey, *Social Science for What?*, chap. 5.

26 Solovey, *Social Science for What?*, 147.

27 See, e.g., James Kilpatrick, "Science Foundation Needs More Scrutiny," *The Sun*, 27 March 1975, A19.

28 James Kilpatrick, "NSF Bureaucracy Hard to Kill," *The Atlanta Constitution*, 20 January 1976, 4A.

29 James Kilpatrick, "Fifth-Grade and Sex," *The Atlanta Constitution*, 1 April 1975, 4A.

30 H. Guyford Stever to Olin Teague, 17 March 1975, NSF Historian 6-27-91.

31 John Conlan to H. Guyford Stever, 15 May 1975, in *National Science Foundation Peer Review, Special Oversight Hearings: Hearings Before the Subcommittee on Science, Research and Technology* (US Government Printing Office, 1975), 21 (hereafter *Oversight Hearings*).

32 H. Guyford Stever to John Conlan, 12 May 1975, NSF Historian, 6-27-205.

33 H. Guyford Stever to John Conlan, 16 May 1975, NSF Historian, 6-27-220.

34 *Oversight Hearings*, 264.

35 Charles F. Brown to Norman Tallent, 12 July 1974, NSF Historian, 6-26-239.

36 *Oversight Hearings*, 182–183.

37 *Oversight Hearings*, 192.

38 *Oversight Hearings*, 208.

39 *Oversight Hearings*, 9.

40 *Oversight Hearings*, 6–8.

41 *Oversight Hearings*, 36.

42 *Oversight Hearings*, 35.

43 *Oversight Hearings*, 141.

44 *Oversight Hearings*, 215.

45 *Oversight Hearings*, 142.

46 *Oversight Hearings*, 502.

47 *Oversight Hearings*, 608–609.

48 *Oversight Hearings*, 470–471.

49 *Oversight Hearings*, 460–461.

50 *Oversight Hearings*, 461.

51 *Oversight Hearings*, 759.

52 *Oversight Hearings*, 780.

53 *Oversight Hearings*, 987.

54 *Oversight Hearings*, 1092.

55 *Oversight Hearings*, 460.

56 *Oversight Hearings*, 390

57 *Oversight Hearings*, 887.

58 The report was a fairly straightforward survey of NSF grant applicants and reflected a general approval of the NSF's procedures. Deborah Hensler, *Perceptions of the National Science Foundation Peer Review Process: A Report on a Survey of NSF Reviewers and Applicants*, prepared for the Committee on Peer Review, National Science Board, and the Committee on Science and Technology, U.S. House of Representatives, December 1976.

59 The NIH director's files contain several memos summarizing the hearings, paying particular attention to instances when NIH review came up during the NSF hearings. See NN3-443-05-006 Box 7, file Comm 2-5-L-2, NIH Director.

60 *Grants Peer Review: Report to the Director, NIH* (National Institutes of Health, 1976).

61 "Decisions by Director, NIH, on recommendations of Grants Peer Review Study Team," 8 February 1978, NIH Director.

62 Mandel, *A Half Century of Peer Review*, 149–151.

63 "Summary Statement and Priority Scores," *NIH Guide for Grants and Contracts* 7 (18 August 1978): 2, NIH Director.

64 "Scientific Peer Review of Research Grant Applications and Research and Development Contract Projects," *Federal Register, Department of Health, Education, and Welfare* 43 (24 February 1978): 7864, NIH Director.

65 "Scientific Peer Review," 7865, NIH Director.

66 George N. Eaves, "Who Reads Your Project-Grant Application to the National Institutes of Health?" *Federation Proceedings* 31 (January–February 1978): 8.

67 Benjamin Lewin, interview with Melinda Baldwin, 24 January 2024.

68 Janice M. Beyer, "Editorial Policies and Practices Among Leading Journals in Four Scientific Fields," *The Sociological Quarterly* 19, no. 1 (1978): 68–88, at p. 76.

69 Benjamin Lewin, interview with Melinda Baldwin, 24 January 2024.

CHAPTER 7

1 Daryl E. Chubin, "Much Ado About Peer Review," *BioScience* 36, no. 1 (1986): 18–21, at p. 19.

2 Eugene Garfield, "Refereeing and Peer Review. Part 1. Opinion and Conjecture on the Effectiveness of Refereeing," *Essays of an Information Scientist* 9 no. 31 (1986): 230–238, at p. 236. The other essays are Eugene Garfield, "Refereeing and Peer Review. Part 2. The Research on Refereeing and Alternatives to the Present System," *Essays of an Information Scientist* 9 no. 32 (1986): 239–248; Eugene Garfield, "Refereeing and Peer Review. Part 3. How the Peer Review of Research-Grant Proposals Works and What Scientists Say About It," *Essays of an Information Scientist* 10, no. 4 (1987): 21–26; Eugene Garfield, "Refereeing and Peer Review. Part 4. Research on the Peer Review of Grant Proposals and Suggestions for Improvement," *Essays of an Information Scientist* 10, no. 5 (1987): 27–33. On Garfield's work and his connections with sociologists and philosophers of science, see Alex Csiszar, "Gaming Metrics Before the Game: Citation and the Bureaucratic Virtuoso," in *Gaming the Metrics: Misconduct and Manipulation in Academic Research*, ed. Mario Biagioli and Alexandra Lippman (MIT Press, 2020), 31–42.

3 Memorandum from Dean George Fraenkel to Dr. William McGill, 10 March 1980. Office of the Provost Records 1953–2006, Box 397, Folder "Tenure 1978–1981," Columbia University Archives, New York.

4 Lionel S. Lewis, "Getting Tenure: Change and Continuity," *Academe* 66, no. 7 (1980): 373–381, at p. 376.

5 Lewis, "Getting Tenure," 377. See also Lionel S. Lewis and Vic Doyno, "The Definition of Academic Merit," *Higher Education* 12, no. 6 (1983): 707–719, at p. 708.

6 Stephen J. Ceci and Douglas P. Peters, "Peer Review: A Study of Reliability," *Change* 14, no. 6 (1982): 44–48, at p. 47.

7 Ceci and Peters, "Peer Review," 48.

8 W. Lee Hansen, "Merit Pay in Higher Education," in *Academic Labor Markets and Careers*, ed. David W. Breneman (The Falmer Press, 1988), 118–119.

9 See, e.g., Barbara A. Lee, "Federal Court Involvement in Academic Personnel Decisions: Impact on Peer Review," *Journal of Higher Education*, 1 January 1985; "The

University World Turned Upside Down: Does Confidentiality of Assessment by Peers Guarantee the Quality of Academic Appointment?," *Minerva* 28, no. 3 (1990): 324–385; Perry A. Zirkel, "Avoiding Litigation in the Tenure Process," *Journal of General Education* 30, no. 4 (1979): 275–281.

10 Ceci and Peters, "Peer Review," 44.

11 David Lazarus to Dr. Susan Garcia, 13 March 1984, in American Physical Society and Physical Review Correspondence 1984, David Lazarus Papers.

12 David Lazarus to Hayes Slaughter, 9 November 1983, American Physical Society and Physical Review Correspondence 1984, David Lazarus Papers.

13 See, e.g., Chih Kawn Chen to S. S. Hanna, 15 March 1983, American Physical Society and Physical Review Correspondence 1983–1984, David Lazarus Papers; Daniel Ira Sverdlik to George H. Vineyard, 26 January 1984, APS and Physical Review Irate Authors Correspondence 1983–1985, David Lazarus Papers.

14 David Lazarus to Hayes Slaughter, 9 November 1983, American Physical Society and Physical Review Correspondence 1984, David Lazarus Papers.

15 "Physical Review Letters Manuscript Referral," 10 August 1982. In American Physical Society: General Subject Files, Box 9, Folder "Physical Review Info 1979–1982," Niels Bohr Library and Archives, College Park, MD.

16 See, e.g., "REFEREE'S REPORT: Chih Kwan Chen, 'Similarities . . .' (LH2019DJ)," American Physical Society and Physical Review Correspondence 1983–84, David Lazarus Papers; "Referee report on 'The W Magnetic Moment in Electroweak Mixing and Composite Models and Amplitude Zeros in $q_iq_j\text{->}W\gamma$,'" 28 February 1983, American Physical Society and Physical Review Correspondence 1981–1983, David Lazarus Papers.

17 Benjamin Lewin, interview with Melinda Baldwin, 24 January 2024.

18 Post's initial letter making the complaint seems to be missing, but we can infer the content from the replies.

19 E. J. Post to David Lazarus, 18 August 1984, APS and Physical Review Irate Authors Correspondence 1984, David Lazarus Papers.

20 E. J. Post to David Lazarus, 19 July 1984, APS and Physical Review Irate Authors Correspondence 1984, David Lazarus Papers.

21 David Lazarus to E. J. Post, 11 July 1984, APS and Physical Review Irate Authors Correspondence 1984, David Lazarus Papers.

22 David Lazarus to G. J. Gabriel, 22 March 1985, APS and Physical Review Irate Authors Correspondence 1983–1985, David Lazarus Papers.

23 A comprehensive list of influential 1980s STS literature would need a book of its own, but a few major works include Harry M. Collins, *Changing Order: Replication and Induction in Scientific Practice* (Sage, 1985); Peter Galison, *How Experiments End* (University of Chicago Press, 1987); Donna Haraway, *Primate Visions: Gender, Race, and Nature in the World of Modern Science* (Routledge, 1989); Andrew Pickering, *Constructing Quarks: A Sociological History of Particle Physics* (University of Chicago

Press, 1984); Steven Shapin and Simon Schaffer, *Leviathan and the Air-Pump: Hobbes, Boyle, and the Experimental Life* (Princeton University Press, 1985); Sharon Traweek, *Beamtimes and Lifetimes: The World of High Energy Physicists* (Harvard University Press, 1988).

24 Daryl E. Chubin and Edward J. Hackett, *Peerless Science: Peer Review and U.S. Science Policy* (State University of New York Press, 1990). The title of chapter 1 was "The Centrality of Peer Review."

25 Sheila Jasanoff, "Peer Review in the Regulatory Process," *Science, Technology, & Human Values* 10, no. 3 (Summer 1985): 20–32; Sheila Jasanoff, *Risk Management and Political Culture: A Comparative Study of Science in the Policy Context* (Russell Sage Foundation, 1986).

26 Deborah Shapley and Rustum Roy, *Lost at the Frontier: U.S. Science and Technology Policy Adrift* (ISI Press, 1985), 99 and 104.

27 See George R. LaNoue and Barbara A. Lee, *Academics in Court: The Consequences of Faculty Discrimination Litigation* (University of Michigan Press, 1987), chap. 5. On Black scholars at US universities in this period, see Jonathan Scott Holloway, "The Black Scholar, the Humanities, and the Politics of Racial Knowledge Since 1945," in *The Humanities and the Dynamics of Inclusion since World War II*, ed. David A. Hollinger (Johns Hopkins University Press, 2006), 217–246.

28 On sex discrimination court cases before 1985, see Margaret W. Rossiter, *Women Scientists in America: Forging a New World Since 1972* (Johns Hopkins University Press, 2012), chap. 2.

29 "How Shyamala Rajender Fought Back Against Discrimination," Chemical & Engineering News, 7 March 2023, https://cen.acs.org/careers/women-in-science/Shyamala-Rajender-fought-back-against-discrimination/101/i9. On the Rajender case, see also Sally G. Kohlstedt and Suzanne M. Fischer, "Unstable Networks Among Women in Academe: The Legal Case of Shyamala Rajender," *Centaurus* 51, no. 1 (2009): 37–62, https://doi.org/10.1111/j.1600-0498.2008.00131.x; LaNoue and Lee, *Academics in Court*, chap. 7.

30 E. R. Shipp, "The Litigious Groves of Academe," *New York Times*, 8 November 1987, sec. Education, https://www.nytimes.com/1987/11/08/education/the-litigious-groves-of-academe.html.

31 Paul Selvin, "Does the Harrison Case Reveal Sexism in Math?," *Science* 252, no. 5014 (1991): 1781–1783, at p. 1782.

32 Paul Selvin, "Harrison Case: No Calm After Storm," *Science* 262, no. 5132 (1993): 324–327.

33 William J. Broad, "Ending Sex Discrimination in Academia," *Science* 208, no. 4448 (1980): 1120–1122, at p. 1121.

34 Broad, "Ending Sex Discrimination in Academia," 1121.

35 Broad, "Ending Sex Discrimination in Academia," 1120.

36 Broad, "Ending Sex Discrimination in Academia," 1121.

37 Philip Goldberg, "Are Women Prejudiced Against Women?," *Trans-Action* 5, no. 5 (1 April 1968): 28–30.

38 Michele A. Paludi and William D. Bauer, "Goldberg Revisited: What's in an Author's Name," *Sex Roles* 9, no. 3 (March 1983): 387–390, at p. 389.

39 Colleen Ward, "Prejudice Against Women: Who, When, and Why?," *Sex Roles* 7, no. 2 (1 February 1981): 163–171, at p. 163.

40 Margaret W. Rossiter, "The ~~Matthew~~ Matilda Effect in Science," *Social Studies of Science* 23, no. 2 (1993): 325–341. Rossiter's title was a reference to Robert K. Merton, "The Matthew Effect in Science," *Science* 159, no. 3810 (1968): 56–63.

41 Douglas P. Peters and Stephen J. Ceci, "Peer-Review Practices of Psychological Journals: The Fate of Published Articles, Submitted Again," *Behavioral and Brain Sciences* 5, no. 2 (June 1982): 187–195, at 189.

42 Peters and Ceci, "Peer-Review Practices," 194.

43 Michael Gordon, "The Role of Referees in Scientific Communication," in *The Psychology of Written Communication: Selected Readings*, ed. James Hartley (Kogan Page, 1980), 263–275, at p. 274.

44 The term "double-blind" has fallen out of favor in recent years because it has been criticized as ableist; I will use the term "anonymous-author review" for the remainder of this chapter.

45 See, e.g., Marcel C. La Follette, "Editorial: On Fairness and Peer Review," *Science, Technology, & Human Values* 8, no. 4 (1983): 3–5.

46 See Wray Herbert, "Blind Reviewing, Hotly Debated, Taking Hold in Humanities Journals," *Humanities Report* 2 (1980): 4–6.

47 Marcel C. La Follette, "Editorial: On Fairness and Peer Review," *Science, Technology, & Human Values* 8, no. 4 (1983): 3.

48 Benjamin Lewin, interview with Melinda Baldwin, 24 January 2024.

49 Cannie Stark-Adamec and Robert Adamec, "Breaking into the Grant Proposal Market: Or, the Trials, Tribulations, Tears, Rage, and Fun (?) Of Acquiring a Quarter of a Million Dollars and Still Being Left Wondering How You're Going to Pay the Rent," *International Journal of Women's Studies* 4 (1982): 105–117, at p. 107.

50 A second reason for the attention on biomedical research seems to have been the recent development of recombinant DNA techniques, which raised concerns about the ethical dimensions of altering DNA. On recombinant DNA, see Doogab Yi, *The Recombinant University: Genetic Engineering and the Emergence of Stanford Biotechnology* (University of Chicago Press, 2015).

51 *Fraud in Biomedical Research: Hearings Before the Subcommittee on Investigations and Oversight of the Committee on Science and Technology* (US Government Printing Office, 1981), 2.

52 *Fraud in Biomedical Research*, 12–13.

53 *Fraud in Biomedical Research*, 167.

54 *Fraud in Biomedical Research*, 179.

55 *Fraud in Biomedical Research*, 60.

56 *Fraud in Biomedical Research*, 177.

57 William Broad and Nicholas Wade, *Betrayers of the Truth: Fraud and Deceit in the Halls of Science* (Simon & Schuster, 1983), 18.

58 Broad and Wade, chap. 2.

59 Broad and Wade, 106.

60 Paul Walter, quoted in Claudia Wallis, "Medicine: Fraud in a Harvard Laboratory," *Time* (28 February 1983): https://content.time.com/time/subscriber/article/0,33009,955142,00.html.

61 See, e.g., John Darsee and Donald Nutter, "Reversible Severe Congestive Cardiomyopathy in Three Cases of Hypophosphatemia," *Annals of Internal Medicine* 89 (1978): 867–870; John Darsee and Steven Heymsfield, "Decreased Myocardial Taurine Levels and Hypertaurinuria in a Kindred with Mitral-Valve Prolapse and Congestive Cardiomyopathy," *New England Journal of Medicine* 304 (1981): 129–135; J. R. Darsee, J. R. Mikolich, N. B. Nicoloff, and L. E. Lesser, "Prevalence of Mitral Valve Prolapse in Presumably Healthy Young Men," *Circulation* 59 (1979): 619–622; John R. Darsee, Steven B. Heymsfield, and Donald O. Nutter, "Hypertrophic Cardiomyopathy and Human Leukocyte Antigen Linkage—Differentiation of Two Forms of Hypertrophic Cardiomyopathy," *New England Journal of Medicine* 300 (1979): 877–882; John Darsee and Robert Kloner, "Dependency of Location of Salvageable Myocardium on Type of Intervention," *The American Journal of Cardiology* 48 (October 1981): 702–710. In the natural sciences, the author who contributed most to the paper's conception and analysis is generally listed first or last in the list of authors.

62 See Barbara Culliton, "Coping with Fraud: The Darsee Case," *Science* 220 (1983): 31–35.

63 See Barbara J. Culliton, "Emory Reports on Darsee's Fraud," *Science* 220, no. 4600 (1983): 936; Culliton, "Coping with Fraud," 32; Barbara Culliton, "Fraud Inquiry Spreads Blame," *Science* 219 (1983): 937. Years later, Darsee's mentor Braunwald would write a reflection on the Darsee affair, admitting, "It is now clear to me that I was unprepared to manage what became essentially a full-scale legal, quasi-prosecutorial investigation, and that as a result the initial stages of the matter could have been handled better." Eugene Braunwald, "Cardiology: The John Darsee Experience," in *Research Fraud in the Behavioral and Biomedical Sciences*, ed. David Miller and Michel Hersen (John Wiley & Sons, 1992), 55–79, at p. 56.

64 For a full account of the Baltimore Affair, see Daniel Kevles, *The Baltimore Case: A Trial of Science, Politics, and Character* (W. W. Norton, 1998).

65 T. Imanishi-Kari et. al., "Altered Repertoire of Endogenous Immunoglobulin Gene Expression in Transgenic Mice Containing a Rearranged Mu Heavy Chain Gene," *Cell* 45 (April 1986): 247–259.

66 See Kevles, *The Baltimore Case*, chap. 2.

67 Doug Stewart, "Interview: Walter Stewart," *Omni* (February 1989): 65–66, 87–94.

68 Stewart, "Interview: Walter Stewart," 90. The *NEJM* paper in question was John Darsee and Steven Heymsfield, "Decreased Myocardial Taurine Levels and Hypertaurinuria in a Kindred with Mitral-Valve Prolapse and Congestive Cardiomyopathy," *New England Journal of Medicine* 304 (1981): 129–135.

69 Stewart, "Interview: Walter Stewart," 90–91.

70 Philip Hilts, "Biologist Who Disputed a Study Paid Dearly," *New York Times*, 22 March 1991, A1.

71 "1996.06.21 DAB1582 Thereza Imanishi-Kari, Ph.D.," accessed 29 November 2023, https://www.hhs.gov/sites/default/files/static/dab/decisions/board-decisions/1996/dab1582.html.

72 Jan DeBlieu, "The Darsee Affair," *Emory Magazine* (December 1983), 8.

73 Walter Stewart, quoted in Jack Anderson and Dale van Atta, "NIH Runs Research on Blind Faith," *Washington Post* (11 April 1988): C15.

74 Bernard Davis, "More Than a Test-Tube Tempest," *Los Angeles Times*, 10 December 1991, VYB13.

75 Robert Pollack, "In Science, Error Isn't Fraud," *New York Times*, 2 May 1989, A25.

76 "Misconduct in Science," 9 February 1983, NN3-443-UD-12D-1 box 78, file "RES 12 Misconduct in Science, 1983–1987," NIH Director.

77 Benjamin Lewin, interview with Melinda Baldwin, 24 January 2024.

78 Richard Saltus, "Baltimore's Legacy: Concern About Oversight of Scientists," *Boston Globe*, 4 December 1991, 67.

79 Stephen Lock, quoted in Marjorie Sun, "Peer Review Comes under Peer Review," *Science* 244, no. 4907 (1989): 912.

80 Drummond Rennie and Annette Flanagin, "Three Decades of Peer Review Congresses," *JAMA* 319, no. 4 (23 January 2018): 350–353, https://doi.org/10.1001/jama.2017.20606.

81 Medical Research Council of Canada, *Celebrating the Medical Research Council of Canada: A Voyage in Time* (Medical Research Council of Canada, 2000), 17.

82 Kathia Serrano Velarde, "The Way We Ask for Money . . . The Emergence and Institutionalization of Grant Writing Practices in Academia," *Minerva* 56, no. 1 (1 March 2018): 85–107, https://doi.org/10.1007/s11024-018-9346-4.

83 Serrano Velarde, "The Way We Ask for Money," 93.

84 "The First International Congress on Peer Review in Biomedical Publication: Guarding the Guardians: Research on Editorial Peer Review," https://peerreviewcongress.org/wp-content/uploads/old_site/pdf/1989/program_1989.pdf, 28.

85 "First International Congress on Peer Review," i.

86 Michael Gordin, *Scientific Babel: How Science Was Done Before and After Global English* (University of Chicago Press, 2015), 180–183.

87 Fyfe et al., *History of Scientific Journals*, chap. 14.

88 "The First International Congress on Peer Review in Biomedical Publication: Guarding the Guardians: Research on Editorial Peer Review," https://peerreviewcongress.org/wp-content/uploads/old_site/pdf/1989/program_1989.pdf, 26.

89 Gordin, *Scientific Babel*, 294.

90 Melinda Baldwin, *Making "Nature": The History of a Scientific Journal* (University of Chicago Press, 2015), 180.

91 Drummond Rennie, "Editorial Peer Review: Its Development and Rationale," in *Peer Review in Health Sciences*, ed. Fiona Godlee and Tom Jefferson (London: BMJ, 1999), pp. 1–13, at p. 4.

92 "Peers Reviewed," *The Lancet* 333 (1989): 1115–1116, at p. 1116.

93 Jon Turney, "End of the Peer Show?," *New Scientist* 127 no. 1735 (1990): 38–42, at p. 38.

94 Turney, "End of the Peer Show?," 42.

CONCLUSION

1 Jocelyn Kaiser, "NIH Will Eliminate Many Peer-Review Panels and Lay off Some Scientists Overseeing Them," *Science*, 7 March 2025, https://www.science.org/content/article/nih-will-eliminate-many-peer-review-panels-and-lay-some-scientists-overseeing-them.

2 "Clare Zhang, Researchers Scramble Following NSF Grant Cuts," AIP, 28 May 2025, https://www.aip.org/fyi/researchers-scramble-following-nsf-grant-cuts; Marcelo Jauregui-Volpe, "New Brief Finds NIH Has Canceled $1.9 Billion in Grants," Association of American Universities, 9 May 2025, https://www.aau.edu/newsroom/leading-research-universities-report/new-brief-finds-nih-has-canceled-19-billion-grants.

3 Max Kozlov, "Exclusive: NIH to Dismiss Dozens of Grant Reviewers to Align with Trump Priorities," *Nature*, 14 July 2025, https://www.nature.com/articles/d41586-025-02221-6.

4 Executive Orders, "Improving Oversight of Federal Grantmaking," The White House, 7 August 2025, https://www.whitehouse.gov/presidential-actions/2025/08/improving-oversight-of-federal-grantmaking/.

5 Russell Vought, "Memorandum to the Heads of Executive Departments and Agencies," Office of Management and Budget, 8 December 2026, https://www.whitehouse.gov/wp-content/uploads/2025/12/M-26-03-Presidents-Management-Agenda.pdf.

6 Alondra Nelson, "Why I'm Resigning from the NSF and Library of Congress," *TIME*, 13 May 2025, https://time.com/7285045/resigning-national-science-foundation-library-congress/.

7 Kathryn Palmer, "NSF Halts New Funding and Caps Indirect Rate Costs," *Inside Higher Ed*, 11 June 2025, https://www.insidehighered.com/news/government/science-research-policy/2025/05/05/nsf-halts-new-funding-and-caps-indirect-rate; Meredith Wadman, "Trump Proposes Massive NIH Budget Cut and Reorganization," accessed 15 July 2025, https://www.science.org/content/article/trump-proposes-massive-nih-budget-cut-and-reorganization.

8 Hannah Daniel, "Commerce Secretary Grilled on NOAA Staffing Shortages," AIP, 11 June 2025, https://www.aip.org/fyi/commerce-secretary-grilled-on-noaa-staffing-shortages; Rob Stein, "Trump Administration Shuts Down EPA's Scientific Research Arm," *NPR*, 20 July 2025, https://www.npr.org/2025/07/20/nx-s1-5474320/trump-epa-scientific-research-zeldin.

9 Rob Stein, "Medical Journals Hit with Threatening Letters from Justice Department," *NPR*, 2 May 2025, https://www.npr.org/sections/shots-health-news/2025/05/02/nx-s1-5374993/medical-journals-hit-with-threatening-letters-from-justice-department; Kristina Fiore, "Medical Journals Get Letters from DOJ," MedPage Today, 18 April 2025, https://www.medpagetoday.com/special-reports/exclusives/115180.

10 Adriel Bettelheim and Tina Reed, "Scoop: Trump Admin Cuts Contracts with Scientific Publishing Giant," *Axios*, 25 June 2025, https://www.axios.com/2025/06/25/trump-cuts-contracts-scientific-publisher.

11 Chelsea Cirruzzo, "RFK Jr. Threatens to Bar Government Scientists from Publishing in Leading Medical Journals," Politico, 27 May 2025, https://www.politico.com/news/2025/05/27/rfk-jr-nih-scientists-medical-journals-jama-lancet-nejm-00371349.

12 Executive Orders, "Restoring Gold Standard Science," The White House, 23 May 2025, https://www.whitehouse.gov/presidential-actions/2025/05/restoring-gold-standard-science/.

13 Michael Kratsios, "Sound Policy Demands Sound Science," *Science* 389, no. 6755 (3 July 2025): 7, https://doi.org/10.1126/science.adz9562.

14 "Publishing Philosophy," Journal of the Academy of Public Health, accessed 15 July 2025, https://publichealth.realclearjournals.org/Publishing Philosophy; "About," Science, Public Health Policy and the Law, accessed 15 July 2025, https://publichealthpolicyjournal.com/about/.

15 Michael Gordin, *Scientific Babel: How Science Was Done Before and After Global English* (University of Chicago Press, 2015), 294.

16 See Aileen Fyfe, Noah Moxham, Julie McDougall-Waters, and Camilla Mørk Røstvik, *A History of Scientific Journals: Publishing at the Royal Society, 1665–2015* (UCL Press, 2022), 530, 573.

17 Loren R. Graham, "Big Science in the Last Years of the Big Soviet Union," *Osiris* 7 (1992): 49–71, at p. 68.

18 "SciELO—Brazil," accessed 19 August 2024, https://www.scielo.br/journal/aabc/about/#instructions.

19 "Journal of Genetics | Indian Academy of Sciences," accessed 19 August 2024, https://www.ias.ac.in/Journals/Journal_of_Genetics/Information_for_Authors; "Journal of the Korean Physical Society," SpringerLink, accessed 19 August 2024, https://link.springer.com/journal/40042/submission-guidelines; "Comptes Rendus de l'Académie Des Sciences," accessed 19 August 2024, https://comptes-rendus.academie-sciences.fr/.

20 "External Peer Review," Medical Research Council, 2018, accessed 12 January 2018, https://www.mrc.ac.uk/funding/peer-review/external-peer-review/.

21 "FAQ: Questions on the Review," Deutsche Forschungsgemeinschaft, 19 February 2025, https://www.dfg.de/en/research-funding/proposal-funding-process/faq/review-process.

22 "Grants-in-Aid for Scientific Research," Japanese Society for the Promotion of Science, https://www.jsps.go.jp/english/e-grants/.

23 Liang Zhang, Justin J. W. Powell, and David P. Baker, "Exponential Growth and the Shifting Global Center of Gravity of Science Production, 1900–2011," *Change: The Magazine of Higher Learning*, 31 July 2015, https://doi.org/10.1080/00091383.2015.1053777.

24 House of Commons Science and Technology Committee, *Peer Review in Scientific Publications: Eight Report of Session 2010–12* (United Kingdom Parliament, London, 28 July 2011), http://www.publications.parliament.uk/pa/cm201012/cmselect/cmsctech/856/856.pdf.

25 William G. Baxt et al., "Who Reviews the Reviewers? Feasibility of Using a Fictitious Manuscript to Evaluate Peer Reviewer Performance," *Annals of Emergency Medicine* 32, no. 3 (1 September 1998): 310–317, https://doi.org/10.1016/S0196-0644(98)70006-X; Fiona Godlee, Catharine R. Gale, and Christopher N. Martyn, "Effect on the Quality of Peer Review of Blinding Reviewers and Asking Them to Sign Their Reports: A Randomized Controlled Trial," *JAMA* 280, no. 3 (15 July 1998): 237–240, https://doi.org/10.1001/jama.280.3.237; Sara Schroter et al., "What Errors Do Peer Reviewers Detect, and Does Training Improve Their Ability to Detect Them?," *Journal of the Royal Society of Medicine* 101, no. 10 (1 October 2008): 507–514, https://doi.org/10.1258/jrsm.2008.080062.

26 For a helpful overview of this issue, see David Shatz, *Peer Review: A Critical Inquiry* (Rowman & Littlefield, 2004), chap. 3.

27 See, e.g., Balazs Aczel, Barnabas Szaszi, and Alex O. Holcombe, "A Billion-Dollar Donation: Estimating the Cost of Researchers' Time Spent on Peer Review," *Research Integrity and Peer Review* 6, no. 1 (14 November 2021): 14:1–14:8, https://doi.org/10.1186/s41073-021-00118-2; Amber Dance, "Stop the Peer-Review Treadmill. I Want to Get Off," *Nature* 614, no. 7948 (13 February 2023): 581–583, https://doi.org/10.1038/d41586-023-00403-8; Colleen Flaherty, "The Peer-Review Crisis," *Inside Higher Ed*, 12 June 2022, https://www.insidehighered.com/news/2022/06/13/peer-review-crisis-creates-problems-journals-and-scholars; Filip Vostal, *Inquiring into Academic Timescapes* (Emerald, 2021).

28 "Retraction Watch," Retraction Watch, 28 December 2024, https://retractionwatch.com/.

29 "Peer Reviewed Articles: Can They Still Be Trusted?," FirstPrinciples, 12 July 2024, https://www.firstprinciples.org/article/is-peer-review-failing-its-peer-review.

30 Larry Wasserman, "A World Without Referees," *ISBA Bulletin* 19 (2012): 7–8.

31 Adam Mastroianni, "The Rise and Fall of Peer Review: Why the Greatest Scientific Experiment in History Failed, and Why That's a Great Thing," *Experimental History*, 13 December 2022, https://www.experimental-history.com/p/the-rise-and-fall-of-peer-review.

32 See, e.g., Owen Dyer, "Trump's 'Gold Standard Science' Rule Would Let Political Appointees Edit Government Science and Punish Scientists, Critics Say," *British Medical Journal*, 2 June 2025, https://doi.org/10.1136/bmj.r1131.

33 Mark Ware Consulting, "Publishing Research Consortium: Peer Review Survey 2015," May 2016, https://assets.ctfassets.net/o78em1y1w4i4/5aqlvrhd07Kcl4X4FjpL5w/49877edd7058156421af83dc30ce0ca7/PRC-peer-review-survey-report-Final-2016-05-19.pdf.

34 The model of approving all papers that are sound science is currently in use at the widely read open-access journal *PLOS One*, where "the editors make decisions on submissions based on scientific rigor, regardless of novelty." "Editorial and Peer Review Process," *PLOS One*, accessed 18 July 2025, https://journals.plos.org/plosone/s/editorial-and-peer-review-process. On *PLOS ONE*, see Martin Paul Eve, Cameron Neylon, Daniel O'Donnell, Samuel Moore, Robert Gadie, Victoria Odeniyi, and Shahina Parvin, *Reading Peer Review:* PLOS ONE *and Institutional Change in Academia* (Cambridge University Press, 2021).

35 "The Urgent Need for a National Plan to Contain the Coronavirus," US Government Publishing Office, accessed 21 May 2024, https://www.govinfo.gov/content/pkg/CHRG-116hhrg41909/html/CHRG-116hhrg41909.htm.

36 Carole J. Lee, "Certified Amplification: An Emerging Scientific Norm and Ethos," *Philosophy of Science* 89, no. 5 (December 2022): 1002–1012, https://doi.org/10.1017/psa.2022.37.

37 See Shatz, *Peer Review*, chap. 1.

38 Remco Heesen and Liam Kofi Bright, "Is Peer Review a Good Idea?," *The British Journal for the Philosophy of Science* 72, no. 3 (September 2021): 654.

39 For a useful review of recent literature on grant peer review, see Kathleen Gregory, Ludo Waltman, and Stephen Pinfield, *Peer Review in Funding Organizations: An Analytical Literature Review* (Research on Research Institute, 2024), https://doi.org/10.6084/m9.figshare.26861680.

40 Simon Swain, "The British Academy Is Trialling a New, Fairer Method of Selecting Its Small Research Grants—Here's Why," 7 September 2022, https://www.thebritishacademy.ac.uk/news/news-comment-the-british-academy-is-trialling-a-new-fairer-method-of-selecting-its-small-research-grants-heres-why/. For more examples of lottery funding, see Jamie Shaw, "Peer Review in Funding-by-Lottery: A Systematic Overview and Expansion," *Research Evaluation* 32, no. 1 (1 January 2023): 86.

41 Apoorva Mandavilli, "How Kennedy's Purge of Advisers Could Disrupt U.S. Vaccinations," *New York Times*, 13 June 2025, https://www.nytimes.com/2025/06/13/health/vaccines-kennedy-cdc.html.

42 This echoes an observation that Alex Csiszar has made: that "moments when the norms and forms of expert communication have been most in doubt are precisely those moments when scientific practitioners have sought—or been forced—to renegotiate their public status within a wider political landscape." See Alex Csiszar, *The Scientific Journal: Authorship and the Politics of Knowledge in the Nineteenth Century* (The University of Chicago Press, 2018), 3.

INDEX

Publisher contact:
The MIT Press
Massachusetts Institute of Technology
77 Massachusetts Avenue, Cambridge, MA 02139
mitpress.mit.edu

EU Authorised Representative:
Easy Access System Europe, Mustamäe tee 50,
10621 Tallinn, Estonia
gpsr.requests@easproject.com

Printed by Integrated Books International,
United States of America